Sex Tour[illegible]

Sex tou[illegible] ight districts [illegible] in raincoat[illegible] *ism*, howeve[illegible] ext. Tourists [illegible] mal restraint[illegible] on- demned[illegible]

Sex T[illegible] rac- tions a[illegible] sive summar[illegible] lary research[illegible]. It looks at [illegible] the implicat[illegible]. It examine[illegible] on sexualit[illegible] s to issues s[illegible] ons wherein[illegible]

The [illegible] sex tourism [illegible] and represen[illegible] the preconc[illegible] the impressi[illegible] ose working[illegible]

Chris [illegible] School, University of Waikato, New Zealand and **C. Michael Hall** is Professor and Head of the Centre for Tourism, University of Otago, New Zealand.

Sex Tourism

Marginal people and liminalities

Chris Ryan and C. Michael Hall

London and New York

First published 2001
by Routledge
11 New Fetter Lane, London EC4P 4EE

Simultaneously published in the USA and Canada
by Routledge
29 West 35th Street, New York, NY 10001

Routledge is an imprint of the Taylor & Francis Group

Typeset in Galliard by Taylor & Francis Books Ltd
Printed and bound in Great Britain by Biddles Ltd, Guildford and King's Lynn

British Library Cataloguing in Publication Data
A catalogue record for this book is available from the British Library

Library of Congress Cataloging in Publication Data
Ryan, Chris
Sex tourism: marginal people and liminalities/Chris Ryan and Colin Michael Hall.
p. cm.
Includes bibliographical references (p.) and index.
1. Sex tourism. 2. Sex-orientated businesses–Social aspects. 3. Prostitution. I. Hall,
Colin Michael. II. Title.

HQ117 .R9 2001
306.74–dc21

00–062801

ISBN 0–415–19509–8 (hbk)
ISBN 0–415–19510–1 (pbk)

Contents

List of figures

Preface

The four S's of tourism 'Sun, Sand, Surf and Sex' are often regarded as being almost synonymous with the supposedly globalising and homogenising nature of international mass tourism. That such a popular representation of mass tourism contains the fourth 's' of sex says a great deal about the images, topographies and realities of the industry and its socio-cultural, economic and political dimensions. Moreover, the inclusion of sex and particular expressions of sexuality is often unquestioned and is accepted by many readers of the language of tourism. For example, the *Let's Go: The Budget Guide to Southeast Asia* notes that 'tourism and the [sex] industry are mutually reinforcing' (Grayman *et al.*, 1996: 183). Tourism undoubtedly promotes specific representations of gender, sexuality and social relationships – a casual glance at travel brochures for most resorts will show 'blue sky' pictures of romantic couples and scantily clad bodies, usually female, around the spa or pool. That tourism commodifies the body has been recognised by some in the academy although not in the industry itself or, if it has been, little concern is expressed unless there is public outrage or condemnation at the more extreme uses of the body (usually female) to promote tourism products.

In consequence, such outrage means that different perspectives are omitted from an analysis of sex tourism. Mullins (1999) presents an alternative scenario based partly upon economic structures in her analysis of sex tourism in Jamaica. She criticises the prevailing form of tourism as one that fails many Jamaicans for producing low levels of personal income, and hence the paradox arises whereby one of the best means for local males and females of achieving higher incomes from tourism rests in being a 'rent a dread' or a female sex worker. Additionally, such earnings have high multiplier effects. Cabezas makes the same argument, writing of the situation in Dominican Republic:

> The circulation of sex workers in the economy brings profits to many, from the transnational hotels and airlines to the small street vendors who sell hair ornaments to be worn at discos. Hotel managers, taxi drivers, business owners, and many other intermediaries traffic the women and usually procure a cut of their earnings. The police, the state, and the local and

> transnational enterprises are all aware that sex has a market value … even while they are proclaiming that prostitution is immoral.
>
> (1999: 105)

Representations of sex and sexualities are therefore integral to contemporary tourism, as are the social and economic structures within which those representations and transactions take place. The actualities of sexuality – the manner in which bodies are used by the tourist – are becoming a more widely acknowledged issue within the public sphere. Issues of rape and sexual assault are also increasingly related to tourism (e.g. see Jones, 1986) while campaigns against sex tourism, and child sex tourism in particular, have also created greater awareness in tourism-generating regions that the relationship between sex and tourism is not necessarily either romantic or beneficial to the destination.

Undoubtedly, there are few issues in the study of international tourism that have become more emotive and prone to sensationalism than that of sex tourism. Discussions on the relationship between tourism and prostitution are frequently speculative while studies of sex tourism often become an act of moralising or sensationalism and frequently suffer from the cultural blindness and sexual taboos which affect attitudes towards prostitution (see Ericsson, 1980). Sex tourism may be defined as tourism where the main purpose or motivation of at least part of the trip is to consummate sexual relations. It might be thought that these relationships are usually of a commercial nature. However, the apparent solidity of such a definition soon starts to fade as the marginalities and states of sexuality start to be explored in more depth. Sex tourism consists of a series of links that 'can be conceptualised as one between a legally marginalised form of commoditisation (sexual services) within a national industry (entertainment), essentially dependent on, but with a dynamic function in, an international industry [travel]' (Thanh-Dam, 1983: 544). Although sex tourism exists throughout the world, it has come to be primarily associated with the travel of tourists, usually male, in the developed world to less developed countries. One of the main attractions is the important cost differential that exists in the provision of both tourist and sexual services in the developing world compared to such provision in the industrialised world. However, such differences in power also exist within the developed world on a regional, national and local basis. Therefore, one of the 'realities' of sex tourism conveyed through the media in the developed world – that it occurs 'over there' – does not hold true. The relationship of sex to tourism is just that, relational. Therefore one of the key issues which the present work addresses is the relationship between visitor and prostitute, subject and object, in a recognition that the reality of sex tourism lies somewhere in between.

Studying sex tourism

Sex tourism has only developed as a legitimate area of tourism studies since the late 1970s. Indeed, Erik Cohen, one of the most widely quoted commentators on sex tourism, noted in 1982, 'It is remarkable ... that a relationship so often casually observed, which provoked so much indignation and exhortation, has generated little interest in serious, unbiased and systematic sociological or anthropological research' (Cohen, 1982: 404). Similarly, in their path-breaking work on prostitution in America, Burley and Symanski (1981: 239) commented: 'The widespread disdain for prostitutes has undoubtedly hindered the ability of many to see or inquire about them, or to receive accurate accounts from informants.' Nevertheless, it must also be emphasised that given the typically informal and illegal nature of sex tourism and tourism related prostitution, and the general unwillingness of politicians and government authorities to even publicly acknowledge its existence, it is extremely difficult to accurately assess the size of the sex tourism industry, its impacts and the accuracy of sex workers and their client's claims (Hirschi, 1962; Wihtol, 1982; Cohen, 1986). For example, the number of sex workers in Thailand has been variously estimated at between 64,886 by government estimates to over one million by academic researchers (see Richter, 1989; Seabrook, 1996; Bishop and Robinson, 1998).

To further complicate studies of sex tourism, the commercial supply of sexual services and prostitution is differentiated in response to the processes of 'capitalist development and conditioning labour relations, demand and supply' (Thanh-Dam 1983: 536). Sex tourism differs in its degree of industrialisation with the formalisation of supply ranging from what may be described as casual, when travellers arrange for their own sexual services, to a packaged form in which services have been pre-booked with hotels and agencies. This latter form appears to have become less common in recent years as public condemnation of sex tourism has grown. However, at the same time, the range of information sources available to sex tourists has perhaps grown, particularly with the growth of the Internet. The supply of sexual services may be spatially and economically segregated between international and domestic markets, and even within international markets differentiation may exist as to the location where client demands are met and the prices paid by purchasers for sexual services.

The sex tourism industry also takes different forms ranging from the production of videos, through to nude dancing in which no direct physical contact occurs and the tourist acts as voyeur, and tourism-related prostitution, of which there are several major forms. First, the casual prostitute or freelancer who moves in and out of prostitution according to financial need. In this situation prostitution may be regarded as incompletely commercialised and the relationship between sex worker and client may be 'ridden with ambiguities' (Cohen, 1982: 411), particularly if the relationship shifts from an economic to a social base. For commentators like O'Connell Davidson and Sanchez Taylor (1999: 42) the distinguishing feature between sex tourism in resorts and the prostitution associated with western red light districts is the 'diverse arrays of opportunities for sexual gratification, not all of which involve straightforward cash for sex exchange'. While these authors describe the processes of mutuality that arise as being based on 'fantasy' it will be argued that the scenarios are in fact more complex and not necessarily totally distinguishable from some of the sex worker–client interactions that occur in non-resort zones. The second, and more formalised form of prostitution, are the workers who operate through intermediaries before visiting clients, for example, brothel workers and the sex workers who operate out of clubs. Since prostitution is generally illegal, prostitutes are often forced to use entertainment establishments such as clubs, bars or other retail outlets in order to operate. On the other hand, brothels are often regarded by authorities as a mechanism for the containment of prostitution in certain locations and the prevention of 'streetwalking' (Symanski, 1981) and possibly also to maximise their own financial returns, illegal or otherwise. A third form of prostitution is that of bonded prostitutes who have often been sold in order to pay debts or reduce loans. In many cases this type of prostitution is a form of slavery, as unlike the previous two forms, prostitution has often not been entered into through an act of will of the individual concerned but instead through the actions of other family members or even through their abduction and kidnapping.

While sex tourism is usually seen in terms of male clients for female sex workers it should also be noted that the reverse has also been recorded. For example, Harrell-Bond's (1978) study of tourism in the Gambia indicated that while female prostitution is common, male prostitution among young Gambians was rampant with middle-aged Scandinavian women openly soliciting the young men with a number then returning to Scandinavia. Similarly, since the early 1980s there has been a small though significant growth in the number of Australian women going to Bali to meet with the local 'beach boys', while Matthews reported that sex tourism in the Caribbean is not solely male-oriented.

> Barbados has earned a reputation in the Caribbean of having a well-developed industry of male prostitution. The industry has a racial dimension in that it thrives on the alleged desire of white female tourists to have sex with black males ... The Canadian secretary syndrome.
>
> (Matthews, 1978: 67)

In addition, there are variations in the extent of child prostitution between destinations. For instance, in a report on child prostitution Rogers (1989) stated that in Thailand child prostitution is 90 per cent female, in Sri Lanka 90 per cent male, and in the Philippines young boys account for 60 per cent of child prostitutes.

Another issue, which complicates the study of sex tourism, is a lack of theoretical and methodological coherence between the different research traditions with an interest in the field. Although Hall (1992) argued that the study of sex tourism potentially united many of the major research concerns of students of Third World tourism, such an observation has not been fulfilled. Different research traditions and researchers, each with their own particular set of values and interests, have been defining the 'problem' of sex tourism in different ways. One of the key points to relate in reviewing the literature on sex tourism is that while sex tourism was of early interest to students of tourism in the Third World (see Richter, 1989), including geographers and political economists, the most substantive research was being undertaken from a tradition that could broadly be described as 'feminist'. Yet, more mainstream social studies typically elect to ignore the social, economic and political significance of sex tourism and the consequent issues of identity, psychology and motivation. Such a situation makes it appropriate to note how research issues clearly have a gendered base (Hall, 1994; Bishop and Robinson, 1998). Indeed, it can be argued that it was only with the emergence of new forms of sexually transmitted diseases, particularly AIDS (e.g. see Clift and Page, 1996), concern over the relationship between sex tourism and international crime (e.g. see Lamont-Brown, 1982), and, to a lesser extent, protests against international sex tourism by women's and church groups (e.g. Cockburn, 1988a, 1988b), that sex tourism began to be seen as an issue which affected men and therefore came to assume greater significance in the social science literature as well as more popular media. Yet, indicative of the confusion that surrounds the subject, it is noted that while the United Nations, in the mid-1990s, moved towards a stance that prostitutes had the same rights under law as enjoyed by others and possessed human and civil rights as sex workers, it nonetheless in December 1997 passed a resolution on 'The Traffic of Women' that sought to prevent sex tourism (Kempadoo and Ghuma, 1999), thereby implicitly denying a right to work. Additionally, Kemapadoo and Ghuma (1999) proceed to argue, much of the debate on sex tourism is also influenced by a concern about child prostitution, a concern that reinforces moralistic positions that are not always the best perspective from which to examine the complexity of adult sex work and its relationship with tourism.

The research process

Initially it was the intention of the authors to examine sex tourism from the classic positivistic stance of non-detachment, but as the evidence mounted the intent and approach changed. In a sense, therefore, the reader will accompany the authors on their own process of discovery, from a theoretical structure of liminality, to what that actually means for the actors involved, to finally arrive at a judgement. It is not expected that all readers will share the conclusions reached by the authors.

In examining sex tourism and issues of sexuality it is, perhaps, impossible for any researcher to be completely withdrawn from the concerns at hand. Indeed, the fact of meeting and talking with sex workers and their clients was found to challenge one's understanding of one's own sexuality and the motivations of undertaking such a work. Within the tradition of ethnographic research in tourism, the role and power of the researcher have been discussed, and it has long been recognised that the very fact of asking a question is not a neutral act. In a talk given at the Albany campus of Massey University in New Zealand on the epistemology of research, drawing on research with strippers and tourists (Ryan and Martin, 2001), this author provided the metaphor of research truths being like an onion. Like many involved in the sex industry, these women have various stock answers for clients as to how they started, what they think, and they select from their repertoire those statements that fit the mood, the time, the client, and perhaps an appraisal of how much money a client may be prepared to give. Over time, a researcher finds that discrepancies become clear in their accounts. As described in the book, a time came when the researcher was asked if he would show his notes. The provision of a poem that encapsulated their statements proved to be one of the catalytic moments in the research process, whereupon it elicited many other 'truths'. In his talk Ryan argued that the research was akin to peeling away the different layers, but it was not to arrive at a core, an elemental truth about stripping, sex and stripper–client relationships. Rather, each layer had its own truth – each story from the repertoire that every sex worker has for a client is a truth of and for that moment. The stories have purposes, they protect the integrity of the sex worker, and reveal only that which is to be revealed for that time, that encounter. But by peeling away the layers, so different 'truths' are engaged upon, but inasmuch as there is a truth, it is the total onion, not the core or the separate layers alone. But equally, as one views the total onion from the outside, the layers beneath the surface are hidden. Consequently it may be possible for any researcher in this area to only engage upon partial truths, from which one can only extrapolate to estimate a more holistic picture. Further, to continue the metaphor, the act of peeling an onion is to continually run the risk of crying. Many of the stories told in the research for this book are not neutral in their content or in their telling; nor are they emotionally neutral. This book contains many fragments of truths derived from interviews with clients and sex workers, and in doing so contains the language of the industry. The authors have sought to construct a story which links these fragments into a structured framework which, perhaps paradoxically, permits different interpretations and unrelated personal histories to still survive. It has been, in many ways, a difficult text to write. This difficulty is re-visited at the start of Chapter 4, which deals with the ways in which tourists and sex workers come to terms with their respective involvement in sex tourism.

Thus, in the act of data collection and writing, the authors became very aware of the research process that is described by Lofland. He wrote that:

> A major methodological consequence of these commitments is that the qualitative study of people *in situ* is a *process of discovery*. It is of necessity a process of learning what is happening. Since a major part of what is happening is provided by people in their own terms, one must find out about

> those terms rather than impose upon them a preconceived or outsider's scheme of what they are about.
>
> (Lofland, 1971: 4)

Part of the learning is a learning about one's self. One issue that was grappled with continuously was that of motive. Was the research simply voyeuristic? Voyeurism has arguably two components – the scopophilic instinct, that is the pleasure involved in looking at another person as an erotic object, and ego libido instincts within which self- identifications and relationships are played out (Mulvey, 1981). As academics the authors were initially drawn to an argument whereby it was obviously the latter with which the research was about. But as sex workers talked, as clients began to say why they engaged in the behaviours that they did, as gays spoke of the pleasure derived from the body, and as simply human beings, the authors had to recognise the role of the scopophilic. As stated in conversation with the first author by one sex worker – such things are simply natural.

As is the case in many academic endeavours, a motive for the research was a wish to know, and to try to offer something more than simply the descriptive anecdotal evidence provided by past researchers with, it seemed, comparatively short periods of exposure to the phenomenon being studied. Consequently, in this book we have sought to provide structures of analysis to help explain and locate sex tourism within the wider context of both tourism and social interaction. Our notes fill many boxes, and by far exceed the number of words used in this book. But publishers' contracts and economics dictated a selection of subjects and evidence, and we are mindful of those things left out. We do feel that this book does significantly differ from any thus far published in the tourism literature. For example, it utilises socio-psychological theorisation in ways not used by previous authors such as those listed above. It very specifically attempts to locate sex tourism within the social settings described because the authors long ago concluded that only by relational analysis can it be understood. Tourism, by definition, is about temporary escapes from roles, responsibilities and everyday social norms – sex workers provide such escapes – yet the nature of the escape, the binding to society of what are defined as marginal people, or perhaps, better described as marginal roles undertaken by people who live in and are part of our society to the more conventional roles of that society can only be understood from the perspective of social and psychological networking within realities of economic and cultural power. In conversation with Catherine Healy, National Co-ordinator of the New Zealand Prostitutes Collective, the first author asked her what had motivated her to continue so long at the post. With a little hesitation she replied that perhaps it was because of a wish to normalise the condition of sex workers. That, too, is perhaps one of the motives for this book; to remove some of the marginality from the lives of some, and yet in cases where women are degraded and exploited, to argue that the marginalities are not those inherent in the status of 'sex worker' but arise from an abuse of economic and cultural power.

Marginal authors

Finally, the authors are both resident in New Zealand. This is important. Arguably, New Zealand as a country is a voice of the periphery (Hall, 1999). Perhaps as part of that marginality, New Zealand has a history of prostitution that differs from that of many other countries. Overseas tourists to our country are sometimes surprised by the fact that 'massage parlours' are legal, are not tucked away on industrial estates or in out of sight rural zones. The business of brothels operates under what may be described as tolerant laws and policing. While this book is international in scope and evidence, nonetheless it is important to note that many of our immediate respondents and contacts come from this environment. This was both a strength and a weakness. A strength because it eased the route to establishing contacts, a weakness because the desperation associated with sex workers in many countries is not part of the daily lives of many with whom we have had long, continuing association. As the metaphor of the onion showed, 'truths' are revealed over time. In New Zealand the authors have discussed the issues with sex workers, clients, and members of the gay community for over five years – by necessity our contacts with those overseas has been much shorter and therefore our understanding is less.

Structure of the book

The book has been structured to first develop a historical perspective to illustrate the modern antecedents of contemporary prostitution and attitudes towards it and sex tourism. The subsequent chapters continue to develop this macro-perspective by de-differentiating texts within contemporary sex tourism to give voice to a number of alternative views. In turn the work moves from the general to the individual as in Chapter 4 individual stories are examined. Chapter 5 then examines how commodities are created from individual concerns about sexual identity by describing the developments of products based around gay and lesbian struggles for recognition. The last two chapters return to the macro-perspective. First, in terms of the modelling introduced in the first three chapters by moving beyond the marginality of the condoned and the tolerated into issues of sex trafficking. Yet even here it is argued, such exploitation is embedded in a structure of poverty and, at times, what can only be termed as evil intent on the part of some who have no pity for the victims of sex trafficking. The final chapter examines the ways in which authorities seek to deal with the extremes of sex trafficking.

Acknowledgements

The final part of a text is a short 'afterword'. Such a topic as this constantly, as indicated above, presents challenges. The authors owe much to many, and there were indeed many to thank for their patience, tolerance and preparedness to answer questions that they did not always want to face. To talk about intimate things that you have not fully analysed yourself at the bequest of an academic researcher is not always easy – acts of discovery can indeed be cathartic and not always welcome. We owe a tremendous debt to a number of people, many of whom to our mind are courageous, strong people coping under conditions of disadvantage that would disable many others. Thus first we would want to acknowledge a debt to the New Zealand Prostitutes Collective, and in particular to Catherine Healy, Michelle McGill and Jo Tassell. Among others are Jacky Stephenson who helped determine those questions and realities that are pertinent to this issue and debts of gratitude are owed to Nicolle Besse, Elisa Howell, Gayna Rowling, Jody Cull, to Sheena, Chrissie, April, Sandy and others of the sisterhood of sex workers. Thank you for your time, your thoughts and shared friendships. We would also want to acknowledge the time willingly given by Sarah Pinwell of WISE (Women employed in the Sex Industry) in the ACT and the Eros Foundation (Canberra) and Dr Susie Kruhse-MountBurton. Susie also acted as a referee for this book, as did Professor Dallen Timothy, and we want to acknowledge the help derived from their comments. Among others of importance was Teressa Bent who willingly listened and commented on some of the outcomes of the catharsis undergone by one of the authors and offered support that showed true tolerance while he worked through his own thoughts. The generous support and honesty of Dr Jeremy Huyton at various stages of the research is hereby acknowledged. Jeremy performed one of the important functions of listening, observing, supporting and not judging that is the mark of real friendship. To Amber Martin, thank you for your time, forbearance and the insights given. Among academic colleagues the support of Professor John Selwood, the late Professor Klaus Alburqueque and the late Dr Martin Oppermann for their written notes and comments about their own research is acknowledged. The help of Dr Julia O'Connell Davidson in providing text from her books before they were published is recognised and thanks given. While we may not agree on the sub-text, the present authors acknowledge the importance of her work in forcing us to more carefully delineate our own thinking. Dr Libby Plumridge of the University of Otago's Christchurch School of Medicine very generously made available some of her published papers and observations and notes from as yet unpublished work of her own. Again, thank you. Among research students who helped at various stages over the past years special thanks go to Rachel Kinder, Emma Robertson, Helen Murphy, Brenda Rudkin, Pinchanok Muangpra and Yingda Fu. Additionally, the help of Debbie Wright in going through the text is acknowledged. The latter two were important in providing Thai and Chinese perspectives that helped our thinking. The willingness of those clients who talked about their relationships with sex workers is also acknowledged. The forbearance of immediate colleagues at our respective universities is also acknowledged, especially those associated with Ethics Committees!

As ever, the actual text is the responsibility of the authors alone.

The authors and publishers would also like to thank the following for granting permission to reproduce images in this work: Blackwell Publishers Ltd, for the figure in *The Psychology of Sexual Diversity* (pp. 63–88), edited by K. Howells, 1984 (figure 4.1); John Wiley & Sons Ltd, for the figure 'The deviant tourist and the crimogenic place' (pp. 23–36), by C. Ryan and R. Kinder in *Tourism, Crime and International Security Issues*, edited by A. Pizam and Y. Mansfield, 1995 (figure 2.2). Every effort has been made to contact copyright holders for their permission to reprint material in this book. The publishers would be grateful to hear from any copyright holder who is not here acknowledged and will undertake to rectify any errors or omissions in future editions of this book.

Note on the text

Each of the authors has taken a primary responsibility for different chapters. Professor Chris Ryan is the sole author of Chapters 1, 2, 3 and 4. While Professor Michael Hall was the main author for Chapters 5 and 7 and wrote the majority of Chapter 6.

Professor Michael Hall wrote much of the initial part of the Preface, while Professor Chris Ryan penned the second part and contributed to the first section. Each, however, read, edited and agreed with the contents of all the chapters.

1 Holidays, sex and identity

A history of social development

If this book is be more than yet another description of sex tourism, illustrated by vignettes of tourist–sex worker interactions, it is necessary to generate a conceptual framework within which such encounters can be understood. The approach selected is an adaptation of what Turner (1982) has termed 'comparative symbology', that is, a comparative study and interpretation of symbols, but one within a wider socio-economic and political framework. It will be argued that both tourist and prostitute are symbols of, and actors seeking, needs generated by a wider social context formed by the modern era ushered in by the Industrial Revolution. Essential to this argument is the contention that both are marginal or liminal people. As marginal people, both tourists and prostitutes have had, at least in Western societies, their roles defined by hegemonies of power, and unless regard is paid to those structures, then at best any description of sex tourism remains but that – a description. Thus this first chapter seeks to establish tourism and that part of tourism known as 'sex tourism' within a historical context and a specific discourse. Within this explanation a number of themes are expounded, implied and hinted at, and in due course throughout this book, these facets will be elaborated and brought to the foreground. First, it will be argued that being a tourist is to occupy a liminal role within a temporal marginality. It will subsequently be argued that this is important in our understanding of sex tourism as in the western world the prostitute is also marginalised. The act of sex tourism can therefore be explained as an interaction between two sets of liminal people – but with a difference. The one, the tourist, is enacting a socially sanctioned and economically empowered marginality, while the second, the prostitute, is stigmatised as a whore, a woman of the night, as the scarlet woman. Yet, as will be described, such stigmatisation is now being challenged, made ambiguous and respectability sought by emphasising the role of female labour within the terminology of being a sex worker. Additionally, any analysis which focuses on women alone as prostitutes is but partial as male strippers and prostitutes are an emerging sector of the sex industry, while homosexual and lesbian holiday markets utilising sex for reasons of relaxation and self-identification are also of growing importance.

Thus, a second theme will be hinted at within this historical review, and that is the process of the holiday as a source of self-identification. Thus, for

MacCannell (1976: 4) the tourist seeks to find the 'structures of modernity' and uses leisure, and holidays, as arenas in which the fragmented modern may recover his or her sense of structure. But the question has to be asked, why is there a necessity in the modern world to use these demarcated periods for such purposes? A neo-Marxist analysis is provided which answers the question by noting the changing modes of production and the commodification of many things in our lives, and the emergence of the tourist as a consumer. Again, links between tourism and prostitution will be noted in this introduction for later analysis. If our periods of escape from the role of worker are commodified, so too, it has been often argued by feminists, sex work is about the commodification (and degradation) of female sexuality (for example, see Dworkin, 1988). Both processes are thus linked to wider social change. But, just as both forms of commodification are reactions to these wider processes of industrialisation – processes which incidentally devalue the non-paid work roles assumed by many in our society – notably those of house-bound women – so too both tourism and sex work contain potentialities to confirm the sense of self that Marx ([1844] 1964) saw as being alienated by modernism. The third theme incorporated within this review attempts to include descriptions of socio-political order in terms of dominant hegemonies that have implications for the development of tourism in general and sex tourism in particular. It will also be apparent from such a viewpoint that both tourism and prostitution possess dangerous forces; forces that, while subordinate to the mainstream of society, by their very presence challenge the norms of the dominant. Their existence continues to represent alternative lifestyles. From a positive stance, in the one case of non-work against work, and in the other, the unwillingness of some women to accept low incomes and thus use their sexuality as sources of income and self-image. However, it is readily recognised that these marginalities also represent the reinforcement of consumerism and the exploitation of women – but, then, ambiguity is inherent in the very nature of liminality.

It may, at first sight, seem strange to begin a book concerned with sex tourism with reference to, however briefly, the wider societal transition from pre-industrial patterns of leisure and work to contemporary situations. But the existing patterns of tourism and prostitution, and their inter-relationships, have common roots in the evolution of our society. It is true that the stigmatisation of prostitutes as fallen women existed before the Industrial Revolution, but it was this period of modernity that legalised, yet made ambivalent, the concept of the prostitute as a *diseased* woman – an attribution arguably made all the easier because the object vilified was female. Yet, simultaneously it was this same period that created a discourse of sex that fantasised and objectified women – it is for example interesting to compare the works of a Zola with their social intensity and descriptions of courtesans or the anonymous Victorian English author of *My Secret Life* with the bawdy tales of a Geoffrey Chaucer, Rabelais or the adventures of a Moll Flanders or Tom Jones. Furthermore, the Industrial Revolution was also a period of empire building and a perceptual construction of an exotic other which engaged the sensual and sexual senses of Victorian

England and the Anglo-Saxon world if not nineteenth-century Europe. In short, many of the attitudes that exist today towards sex, tourism and sex tourism have their historical antecedents in movements of the last two hundred years, and to ignore these is to only partially understand how we have reached the current position and why within prostitution and tourism there lie responses of coping with and challenges to the status quo. It is a status quo born of complex socio-political-economic power structures, and thus any summary of a holistic over-view will be selective, incomplete, but hopefully sufficient to show that the themes of marginality, self-identity and economic-political hegemony are important in any analysis of sex tourism.

As tourists and prostitutes are both real in themselves, but symbolic of the consequences of industrialisation and the later commodification of services and values associated with postmodernism, it may be advantageous to define some terms. As Turner (1982) notes, comparative symbology is narrower than semiotics but wider than symbolic anthropology. Semiotics incorporates three branches of study of the nature and relationships of signs in language, namely:

1. *Syntactics* – the formal relationship and organisation of signs and labels within language, the syntax of language.
2. *Semantics* – the relationship of signs and symbols to the things to which they refer.
3. *Pragmatics* – the relationship of signs and symbols with their users.

The concern of this analysis is not with the technical aspects, the syntactics involved with symbology, but with the relationships between users of touristic and sexual services. Its concerns lie with the symbols they represent within a wider society, the way in which they are represented by power structures within that society and the ways in which interactions between prostitutes and tourists are played out within these relationships of power and symbolism. It has already been stated that sex tourism within this analysis is perceived as an interaction between two marginal groups, tourists and sex workers, and thus this analysis develops that commenced by Ryan and Kinder (1996a, 1996b). They concluded that conventional descriptions of deviancy were not sufficient to explain either the behaviour of the tourist who sought the services of prostitutes, and neither do such concepts fully explain the role of the prostitute in contemporary society.

What then is meant by liminal or marginal people? Liminal people are threshold people existing betwixt and between. They exist in an ambiguous position between 'positions assigned and arrayed by law, custom, convention, and ceremonial' (Turner, 1969: 95). Turner notes that liminal entities may be disguised, wearing only a strip of clothing, or even going naked among pre-industrial societies, thereby signing that they have no status, property, or insignia. How pertinent is it that the tourist sunbathes with little clothing, and the discerning symbol of the prostitute is the short, figure-hugging mini-skirt and low-cut dress? Boyle describes the 'work gear dictated by clients' demands and the environment'. Writing about 'Maria' she says:

> If she is street walking, she opts for either a red leather mini-dress and thigh-length boots or black four-inch stiletto-heeled shoes, stockings and suspenders. They are carefully positioned to show just below the dress or a black stretch skirt which barely skims her thighs. Accompanying the skirt will be a lacy, gravity-defying basque which attempts to encompass her magnificent bosom.
>
> (1994: 131)

In an earlier period Mayhew ([1851] 1999: 416) draws our attention to the power of undress, dress and their significance as signifiers of the marginal. He cites one prostitute who says of herself, 'I have good feet too, and as I find they attract attention, I always parade them. And I've hooked many a man by showing my ankle on a wet day.' In the next paragraph Mayhew refers to 'black silk cloaks or light grey mantles – many with silk paletots and wide skirts, extended by an ample crinoline', but just as Odzer (1994) was to note a century and half later, a hierarchy existed wherein at the bottom in the Haymarket were 'wornout prostitutes or other degraded women, some of them married, yet equally degraded in character (Mayhew, [1851] 1999: 417).

Lodge (1992), in his novel, *Paradise News*, describes tourists dressed in shell suits and shorts – signs of being a tourist. Both representations are concerned with a display of the body and nakedness or near nakedness are involved. Yet, the realism of the display is that it is not the appearance of glamourised bodies seen in pin-up calendars. The short, the fat, and the skinny are revealed and both tourists and prostitute have to come to the truth of their bodily appearance even while both hide and display the body at the same time. Annie Sprinkle's list of reasons as to why whores are heroes contains, at numbers 13 and 37, the reasons 'Whores wear exciting clothes' and 'Whores are not ashamed to be naked' (Sprinkle and Gates, 1997). Thus again the ambiguity, the dialectical tension of opposites is evident within this marginal behaviour. In conversation with the first author, Michelle, a former sex worker in New Zealand, commented that there were many ambiguities in the dress worn by prostitutes. The point of dressing was to enhance the body, but one dresses to become naked, but naked in a flattering environment. There is, she commented, 'the sensuality of undressing – one dresses to undress'.

Within marginal groups, states Turner (1974), there exists a sense of *communitas*, of homogeneity and comradeship – they possess an area of common living. This is almost literally true for sex work. Areas of prostitution, termed 'red light' areas, offer mutual support systems for their inhabitants. Women on the street take down the number plates of cars that their fellow women get into in case there is any trouble. Clients, pimps, workers and drug pushers mark out territories so as to more easily sustain social relationships. It is consistent with the theories of liminality that the safety of such areas is fragile and what, at least for the client, they offer is the opportunity for anonymity (Ryan and Kinder, 1996a). So too with tourism. As is noted below, tourists are found in purpose-built tourist spaces. It can be objected that these relate solely to mass tourism, but in response two viewpoints may be held. One may hold to the position of Boorstin (1963) that these are the epitome of tourism, and not to be present at such a place means that one is a traveller. Boorstin defines the modern experience of tourism thus:

> [But] the experience of going there, the experience of being there, and what is brought back from there are all very different. The experience has become diluted, contrived, prefabricated.
>
> The modern American tourist now fills his experience with pseudo-events. He has come to expect both more strangeness and more familiarity than the world naturally offers. He has come to believe that he can have a lifetime of adventure in two weeks, and all the thrills of risking his life without any real risk at all.
>
> (Boorstin, 1963: 88)

For Boorstin these facts described the very character of modern tourism – from the perspective of an analysis of tourists as marginal people it seems to describe an essence of liminality. The tourist exists in an irregular world that is both strange and familiar. At the other extreme even Cohen's (1979) 'drifters' are notable by their style and by their tendency to drift towards certain roles and places. Travellers too are marginal persons, members in but not part of the crowd. For Bauman (1994) and the concept of tourist–traveller as *flâneur*, the tourist is one who

> [goes] for a stroll as one goes to the theatre ... (in the crowd but not of the crowd), taking in those strangers as 'surfaces' – so that 'what one sees' exhausts 'what they are', and above all seeing and knowing them episodically ... rehearsing human reality as a series of episodes, that is events without past and with no consequences.
>
> (1994: 27)

Marginality assumes further importance by reason of temporality and transition. The tourist assumes the role of non-worker. The holiday trip is characterised by stages of preparation, absorption into the tourist role and then re-entry to the mainstream world. Each stage may be characterised by small rites – the process of checking in for flights, the purchase of holiday clothing, and the development of the photographs after the trip. The prostitute occupies other roles besides being a 'working woman'. She is mother, student, partner. As noted she dons the costume of the night – an act of ritual wherein the dress conveys power. Kasl (1989: 154) writes '[the] addiction part is the ritual of getting dressed, putting on makeup, fantasising about the hunt, and the moment of capture. To know that you go out there and they would come running. What power!' Thus Kasl demonstrates another paradox about the state of marginality, and that is the dialectic between powerlessness and power. The woman who is disdained is the woman with power over men. Likewise the tourist as hedonist, someone apart from the Puritan work ethnic, is pampered by an industry that recognises such hedonism as the reward of work. The non-worker, normally a position of powerlessness, possesses power over the worker. The reality, as Marx would have pointed out, is that such power is exercised by cash. But the differences of each respective marginality, sanctioned or condemned by society, again become evident. The working girl exercises a power to earn cash; the tourist exercises a power due to the possession of money. Hence the possibility that the tourist may purchase the services of the prostitute or other sex workers. Therefore it is not surprising, given the hedonistic nature of tourism, that in many places the spatial areas of both tourist and sex worker overlap, and many hotel managers can bear witness to the fact that their premises might be regarded as both holiday accommodation and brothel. In some parts of the world the overlap becomes obvious and explicit. Amsterdam's red light district and the *soi* of Patpong are tourist attractions for clients and on-lookers alike.

To travel has been variously defined as an act of pilgrimage, escape, a search for adventure or for self. The rewards it offers are generally intangible, and those things that are purchased, as souvenirs, are mementoes designed to evoke memories. And arguably it is not so much the place that is evoked, but rather what was done at that place – the people one met, or the relationship between self and place through processes of evaluation – whether of the aesthetic values or events. In short, as humans, we can only experience place through the filter of our own selves. Urry (1990) has drawn attention to the role of the tourist gaze, but the perceptual is related to the corporeal. The role of the body is being recognised in studies of tourism. Our own bodily comfort or discomfort engages senses that also help to shape perception of place and activity. Describing the experiences of a student tourist group in Sarawak, Markwell describes how fatigue caused by climbing Mount Kinabalu the previous day, shaped activities and thus consequent memories and evaluations of a visit to an orangutan sanctuary. Thus one of his respondents, Jane stated:

> I was really disappointed that we didn't get into it as much, like we'd really been looking forward to it, and I just remembered the day that we went there it was really hot and all we could think of was just getting to the cafeteria and like getting a drink and actually once we got there it was like, 'Oh, it's too hot to go back outside', and I really regret it because a lot of people went further up and they had contact with them [orangutans] and everything and we didn't and that's one thing I've really regretted.
>
> (Markwell, 1998: 191)

Similarly, in an examination of eco-tourism in the heat and humidity of the 'Top End' of Australia, Ryan *et al.* (1999) contend that eco-tourism, like all forms of tourism, is concerned with spectacle. It is, they argue, designed to focus on culturally acceptable sights, but is done in such a way that the sight is elevated above the other bodily senses through the provision of shade in wind accessible places, thereby creating an experience of place that is a managed experience of raw nature offered by the climate. The elevation of the body as a source of pleasure is thus another theme that must be explored within tourism in general and sex tourism in particular. Sun bathing brings not only the possibility of a suntan, but also the sense of sun upon the skin, a warmth and a pleasure. From sensuality to sexuality is a stepped procedure. And from sexuality to intercourse is but a further graduation based on payment. From the warmth of sun to the warmth of a massage to the body massage of Spalding Gray's bubbling lotions as described in his 1989 one-man show based on his text, *Swimming to Cambodia* (1985), to the act of intercourse – isolated to the context of the body, the relationships seem obvious. Gray describes this process with the Thai massage parlour as:

> You go down into this small room and for a little bit of money you take off all your clothes and she stays dressed, and you get a mild, tweek-tweek massage … A little more money and you get a hand job. A little bit more money and you get to fuck her. A little bit more money and you get the supremo-supremo … the body-body massage … And she gets on one side of the room and runs and hops on top of you and swiggle-swiggle-swiggle, body-body-body, and you slide together like two very wet bars of soap.
>
> (Gray, 1985: 42)

Yet even within this very narrow conceptualisation the role of sight has pre-eminence, at least in the early stages of any action. The sex industry is one of presenting images, and the choice of sex worker by the client is based upon sight as much as any other sense. Gray chooses his partner based on looks. 'Sarah' – a sex worker in Blenheim, New Zealand, describes the fact that escort work is easier than working in a massage parlour because there, in the parlour, one competes with other women for the custom of the 'gentleman' (Spectrum, 1998). Sensuality and sexuality – related but separate entities struggle for space in any understanding of tourism, prostitution and sex tourism. This issue of sensuousness versus sexuality is yet another theme to be disentangled within a conceptualisation of marginality.

Whether pleasant or unpleasant, whether uplifting, hedonistic or degrading, what remains of any tourist trip is an experience. Regardless of whether that experience fades with time as memories dim, or whether it remains real, re-interpreted and re-evaluated so that it may become a different thing, at some time it was thought to be important. It was sought out, or if it happened by chance, either prolonged or shortened. The tourist trip is full of encounters which involve us, the tourists, as actors and directors in the play that is the tourist trip – a play staged by a tourist industry that continually advertises the opportunities for 'experience'. That this is the case should not surprise. Experiences possess importance to people. They form memories which help to pass time in the future, to help us survive hard times, and from memories we create expectations as to possible outcomes from future actions. But tourist memories are arguably a special category of memory, as holidays are specially demarcated special times. For Hollinshead (1999: 11) 'the tourist gaze', as described by Urry (1990), 'helps conduct people to all manner of new ecstacies in their travels and enjoyments'. The usual routine is set aside, and often travel to other places is involved. By this very fact they become marginal periods in our lives of potentially great importance. During these periods we live on the perimeter of our normal roles. They are not entirely divorced from those roles and we may carry into our holidays some of our normal roles, for example, as partners or parents. We carry into our holiday periods our skills, be they social skills or skills associated with our occupations. But primarily we leave behind occupational roles, social frameworks that guide or constrain our actions, and enter a period which society sanctions as a period of release from normal responsibilities.

From this perspective sex tourism is located firmly within the wider discourse of tourism. The sex industry presents spectacle in its portrayal of sexual adventure – the modes of dress as already noted are designed to appeal, to excite. Its literature promises ecstasy and finally, orgasm. Yet, within the sex industry lies ambivalence for the exotic dancers of striptease bars offer fantasy but no sexual intercourse, and dancers perform within circumscribed roles on the part of clients (Hanna, 1998; Ryan and Martin, 2001).

It is not for nothing that writers like Shields (1991), Rojek (1993) and Ryan (1997) have alluded to medieval periods of disorder when discussing holidays. During such periods settled order was set aside, the jester became king and hierarchies were challenged, albeit within a temporal framework where such disorder was temporary. But as Rojek and others emphasise, it was a specific form of disorder – it was the wantonness of carnival – lewd, rude, sexual – it was the farts and not the thoughts of radical rebellion against accepted order that characterised such periods. Thus the challenge to accepted order was made safe and earmarked as being an 'other'. The very licentiousness of the form of assault upon normality undermines and thereby makes safe its challenge by de-politicising the nature of the challenge. This phenomenon continues today. Lawrence (1982) describes the Doo Dah Parade in Pasadena as a parade set up as a ritual of rebellion against the Rose Parade. This latter parade, which is organised, sponsored by business, possesses floats of flowers and innocuous smiling beauty queens. Doo Dah parodies it and its blandness. Consequently Doo Dah is held by some residents to be tawdry, cheap and tasteless, but as a moment of misrule and disorder it is ambiguous in that it too is organised and became as successful a tourist attraction as that which it sought to parody. Yet it is of interest that within this context that the body, which was addressed previously in terms of sensuousness adopts another theme in this context – it is the body of orifice and lewdness. The symbolism of the body as caricature has also been noted in tourism. For Goldsmith, a female commentator of the political right, cites her objection to the homosexual displays of Sydney's Mardi Gras on the basis that:

> Some drag queens are an undisguised and vicious parody of woman-hood: their grotesque make-up and clothing and their grossly exaggerated breasts and mannerisms turn women into a joke ... Again, some of the bondage in the Mardi Gras is a worry, when it displays, for example men leading women along by chokers around their necks. In a society where women are demonstrably less than equal and are subject to domestic and sexual violence, such images reinforce negative stereotypes and show violence against women as acceptable.
>
> (1996: 78)

The carnival exists to defy and challenge 'normality'. So too, it can be argued, is the case with holiday periods, but in this case the defiance of normality has been tamed and commoditised. Holidays are periods of relaxation, and thus by their nature a statement that there exists a world other than work. Yet the lifestyle represented by holidaying is itself a de-politicised lifestyle – it is a reward for work and thus a creation of the process that created the discipline of work in the factories of the last century. The holiday is itself an invention of the modern world emerging from the industrial revolution. Rybczynski (1991) describes the history of 'The Weekend' as a response by the dominant classes in Victorian Britain to meet the needs of both God and Mammon. Describing 'Saint Monday', he writes:

> Saint Monday may have started as an individual preference for staying away from work – whether to relax, to recover from drunkenness, or both – but its popularity during the 1850s and 1860s was ensured by the enterprise of the leisure industry. During that period sporting events, such as horse races and cricket matches, often took place on Mondays, since their organisers knew that many working-class customers would be prepared to take the day off. And, since many public events were prohibited on the Sabbath, Monday became the chief occasion for secular recreations.
>
> (Rybczynski, 1991: 43–44)

A combination of industrialists and evangelicals combined to create the weekend where a half-day holiday was provided on the Saturday. Theories of social interaction between popular practice and the interests of the ruling classes also help to explain the emergence of the modern holiday. Cross notes reactions to legislation to paid vacation time in the United Kingdom thus:

> Employers' reaction to the paid vacation was not nearly as hostile as it had been to the shorter workday. The vacation was a prerequisite which could raise work discipline (by denying it to workers who failed to remain on the job for at least a year or to workers with bad records of absenteeism). A

> paid vacation would 'implant in the minds of the employees that they were actually part of the business'.
>
> (1989: 602)

Cross also argues that employers of that period also gained from being able to lay workers off during seasonally slow periods. They also benefited by conceding only forty hours of the working week for an extra week of holiday compared to 400 hours which would be lost by a movement from a forty-eight-hour week to a forty-hour week.

Ryan (1997) has described holidays as periods of potential re-discovery, of relaxation, of sanctioned escape but has omitted to assess the nature of the economic relationships within which the holiday emerged, even while describing its role as a continuation of carnival. Like Doo-Dah, the holiday has become an example of consumerism, not a challenge. While at the level of individuals holidays can be a source of cathartic experience, at a macro-social level holidays are conformist and themselves a source of further consumerism and industry creation.

To understand the nature of sex tourism requires an examination of the relationship between the body, sex and self-identity, the commoditisation of these periods of time and the liminal nature of these experiences, places and the tourist role. But in addition to these themes, and to make sense of how they become apparent, attention must be paid to the wider socio-economic-political context which determines how these forces shape holiday experiences. It should not be forgotten that tourism is big business – a whole industry, global in scope, exists to transport, feed, house and entertain tourists. But it is, arguably, an industry derived from, and shaped by other industries. It has been noted that holidays are a modern phenomenon. They are a necessity created by modes of production, which, from the 1880s, created the production conveyor belt, a system unknown to pre-industrial society. Industrialisation created a scenario where we work five or six days to relax, work free, on the remaining day(s) of the week, an apparent justification of the Marxian notion of man as the appendage to the machine. For Marx and Engels, the success of the bourgeoisie, meant that:

> It has pitilessly torn asunder the motley feudal ties … And has left no other bond between man and man than naked self-interest, than callous 'cash payment' … It has resolved personal work into exchange value and in place of the numberless indefeasible chartered freedoms, has set up that single, unconscionable freedom – Free Trade. In one word, for exploitation, veiled by religious and political illusions, it has substituted naked, shameless, direct, brutal exploitation.
>
> (Marx and Engels, [1844] 1993: 47)

Marx and Engels argued in the *Economic and Philosophic Manuscripts of 1844* that the result was an estranged labour, by which man became objectified as a unit of labour. Thus man was turned into 'a being *alien* to him, into a *means* to his *individual existence.* It estranges man's own body from him, as it does eternal nature and his spiritual essence, his *human* being'. It further led to man being estranged from man and 'If a man is confronted by himself, he is confronted by the *other* man' (ibid.: 63).

The modernism of the Industrial Revolution created a need for and an ability to buy periods of sanctioned escape for purposes of relaxation. It commoditised relaxation in order to allow people to be better workers through a recognition of the fact that people cannot work continuously and thus, to be more productive, required holidays. By offering these periods of escape, by encouraging escape to other places, by validating such behaviour as being educational, it created an additional product – the holiday – which too could take its place in the portfolio of things to be purchased by a growing consumer class.

For Marx a sense of identity and the alienation of that identity were a consequence of an economic system wedded to capitalist systems that valued objects, and people, solely in terms of cash. The echoes of that process continue today. For Bishop and Robinson (1998) the capitalist system and the classifications of 'other' that are created by fragmentations based on hegemonies of European-American male-based cultures are important in explaining the emergence of sex tourism in Thailand. There is, they contend, a standard 'Objective' narrative about the Thai sex industry. As Phongpaichit and Baker note, it commences with:

> the decay of local communities leading to large-scale migration of rural girls (and later, also boys) to work in prostitution for the U.S. soldiers, for an increasingly prosperous urban market, later for the tourist trade, and finally as an export commodity.
>
> (1995: 75)

This is, as far as it goes, an accurate summary, but it fails to ask a number of important questions that are both economic and socio-cultural in character. In 1971 Robert MacNamara, as head of the World Bank, visited Bangkok and initiated a series of subsequent meetings that generated plans whereby tourism was identified as a major component of the Thai economy. Truong (1990) noted that there existed a political motive for such plans. The unstable nature of the Indochina political structures of the time, that is the threats to American hegemony in the region in the late stages of and the post-Vietnam War period, meant that it was necessary to bolster Thailand's economy as that country was generally friendly to American aspirations in containing both communism and uncertainty in the region. The structures created by R&R facilities, an important economic resource, would also fail without alternative sources of demand. Bishop and Robinson (1998: 99) propose a thesis that a cynical rationality took place. Realising that because of then existing fare structures and resultant high costs of travel to Thailand, any tourism development was unlikely to attract family-based tourism, the World Bank and the Thai government colluded in a tourism policy whereby it was inevitable that a tourism based on prostitution could not but help occur. Thus was set in progress a chain of action whereby an economy was created based on sex tourism which continually re-confirmed the importance of massage parlours and bars – and in that re-confirmation failed to provide alternative forms of development, tourism based or otherwise.

It is possible to interpret this process as one whereby a marginal spatial location, Thailand, gained an importance because the economic interests of the United States required it to support a friendly country to stabilise a geo-political zone after a military defeat. Allied to these lay conceptualisations of an exotic other derived from a recent past born of popular literature and Hollywood. Bishop and Robinson (1998) argue the notions of Thai women as exotic, pliable, sexually innocent but fun-loving people emerge from a nineteenth-century conceptualisation of and fascination with harem life, especially as described by popular works like those written by Anna Leonowens *The Romance of the Harem* (1991). In 1862 Leonowens was employed by King Mongkut of Siam. Her early life, spent in India, and subsequent travels in the Middle East as the unchaperoned companion (at the age of 15) with a Reverend George Percy, had brought her into contact with the Muslim harem. Bishop and Robinson argue that these perceptions of the Muslim harem were applied by Leonowens to the Nang Harm or Royal Harem of the Siamese court. Through a series of reprints under different titles, *The Romance of Siamese Harem Life*; *Siamese Harem Life*; *Siam and the Siamese*, Leonowens represented the Siamese Harem as a location of the oppression of females, a place of slavery and of sexual purpose. However, the messages of women's rights were submerged within a titillation of popular taste. Morgan (1991) argues that these messages were also lost due to the non-sanctioned source of the message (Leonowens was female, of mixed blood and not of the upper class). Of more importance was the interpretation of the book by its audience predisposed to a perception of the Far East as an exotic land with different norms of sexual behaviour. The significance of Morgan's comments is that, within the processes of hegemonies being established by a newly industrialising society, popular writing about harems, an exotic subject, transposed to a new and possibly even more exotic location by a woman who herself could be considered as an adventurer – all these factors conspired to reinforce notions that 'out there', beyond the boundaries of everyday society, there existed a different way of conducting male–female relationships; one where women were subordinated to male pleasure. Furthermore, free from a Judaeo-Christian tradition, those women who were providing sexual services lived in a society where intercourse with other than a married partner was not seen, in itself, as immoral. It is possible to imagine that for the Victorian gentlemen personified by the author of *My Secret Life* (Anonymous, cited by Foucault, 1990), such revelations could not be other than attractive. Foucault ([1976] 1990) draws our attention to the fact that the Victorian period was not a period of sexual repression, but of sexual discourse in ways not previously undertaken – a discourse of detail, made moral and acceptable by scientific inquiry. Thus Foucault writes: 'A censorship of sex? Rather there was installed an apparatus for producing an even greater quantity of discourse about sex, capable of functioning and taking effect in its very economy' (1990: 23).

Certainly, within nineteenth-century London, the economic implications of prostitution were significant. It was, for women, a major source of employment. Mayhew ([1851] 1999) estimated that in London, in the 1850s, there were approximately 80,000 prostitutes. By comparison Matthews (1997) estimated that in London, 635 women work 'off the street', 640 in private premises, 2,220 in massage parlours or sauna clubs, 1,260 as escorts and 500 in hostess clubs. He additionally estimated that weekly they entertain 80,240 clients for an annual turnover of £194 million. In short, prostitution was more embedded in London society in the nineteenth century than it is today if the criterion is solely that of numbers of both sex workers and clients.

For Bishop and Robinson, then, based on these foundations of Victorian sensuality, the elements of Leonowens' work that percolated through to *Anna and the King of Siam*, and the subsequent musical, *The King and I*, were those to which a wider populace were responsive. Allied with other sources of representation that are discussed by Bishop and Robinson, an exotic 'other' of friendly, relaxed, smiling people lay waiting for the tourist. They note 'Tomorrow will do, the Thais seem to say; relax, nothing is *that* serious, life is amusing, why not enjoy it?' (Bishop and Robinson, 1998: 33). And of course, if nothing is that serious, neither is sex.

The exoticism of the Far East and South Pacific is commented upon by many writers. Thus Theweleit (1987) notes the derivation of the myths of the South Seas when the image of the South Sea Maiden 'began to construct the body that would constitute a mysterious goal for men whose desires were armed for an imminent voyage, a body that was more enticing than all the world put together' (1987: 296) while Rose (1993: 94) draws attention to the 'sexual, fertile, silent and mysterious Woman with a gorgeous, generous, lush Nature' painted by the artist Gauguin in his representation of Tahitian women. For feminist writers like Rose, geographers have failed to understand or recognise the 'sensual topography of land and skin' which is, she argues, mapped by a mainly white male heterogeneous gaze. Given this historic context of industrialisation fragmenting the world-view into relationships of the particular in which the nineteenth-century Anglo-Saxon mind ascribed roles while secure in a confidence born of masculinity and economic power, it can be argued that it is of little surprise that because of the marginality of location, peoples and the exoticism ascribed to those people, sex tourism has become a sustainable force at the commencement of the twenty-first century. The historical shadows from the past are indeed long ones.

Those shadows, in the case of sex tourism in places like Thailand, are significant, argues Manderson, when the exoticised oriental female entity that Thailand became in the Western mind, itself becomes a scenario and a source of hedonistic self-discovery through sexual intercourse as displayed in the film *Emmanuelle* (1974). Manderson (1997) argues that the film was *not* about Thailand – she writes 'Thailand is simply the backdrop for a European pornographic fantasy' (1997: 136), but 'the film fed into an emerging image of Thailand as sex-haven, and this was not accidental' (ibid.). In its messages of self-discovery by setting aside repression of sexual desire and experimentation, and siting that hedonistic sexual adventure in Thailand, the film reinforced notions of an erotic location and certainly did not harm the business interests of the emerging Thai sex tourism industry. Manderson goes on to trace an evolving sexual imagery of Thailand from the *King and I*, via *Emmanuelle* to the almost total commoditisation of Thai women as sexual objects in O'Rourke's film *The Good Woman of Bangkok* (1991). By 1991, however, the designation of Patpong as a sex tourism destination was well known, and O'Rourke's film is a recognition of the ambiguity of both sex tourism and the director as sex tourist. As Manderson comments, its analysis is akin to those of published sociological works on prostitution in Thailand. It reveals the contradictions of modern complicities as being an inheritance of nineteenth-century certainties. The sequence from the *King and I*'s representation of the governess as a civilising influence to the discourse of O'Rourke's film represents the dissolution of modern certainty into a postmodern recognition of multiple truths. To Manderson's text might be added the example of *Ms Saigon*, a further representation of Western perspectives of sexuality located in South-east Asia.

This analysis, a mixture of neo-Marxian, Freudian and Foucauldian thought, and incomplete though it may be in its contemporary interpretation of late capitalistic modes of production, begins to explain the growth of modern holiday taking and its relationship with the sex industry. To reiterate, first, as is evidenced from countless reports of the Poor Law Commissioners in the United Kingdom, and as described by Thomas Carlyle in *Past and Present*, ([1843] 1965), social life was divided and fragmented into periods and roles dominated by work and occupational status in the late nineteenth century and early twentieth century in Europe and Northern America. Fragmentation creates alien others, people not within specific social circles, activities that are not approved by the dominant *mores* in society and, with the growth of imperialism and the empires of the European nations in Africa and the East, exotic other places (Douglas, 1996). It can be argued that this process of demarcation was tinged with the attributing of moral values which served to reinforce the superiority of those making judgements. Such judgements extended to the discussion of gender, and the previous holistic assessments of sexuality and its role within the natural order of things were cast aside by many within the newly industrialising age. At its most extreme this was demonstrated by attitudes towards recently subjected indigenous peoples. Ann Cameron provides evidence of this from a recent past in quoting evidence from Indian women who belonged to a matriarchal, matrilineal society in Western Canada. Thus she writes:

> [T]he priests had to be content to take the girl children. Instead of being raised and educated by women who told them the truth about their bodies, the girls were taken from their villages and put in schools where they were taught to keep their breasts bound, to hide their arms and legs, to never look a brother openly in the eyes but to look down at the ground as if ashamed of something. Instead of learning that once a month their bodies would become sacred, they were taught they would become filthy. Instead of going to the waiting house to meditate, pray and celebrate the fullness of the moon and their own bodies, they were taught they were sick, and must bandage themselves and act as if they were sick. They were taught the waves and surgings of their bodies were sinful and must never be indulged in or enjoyed.
>
> (Cameron, 1988: 61–62)

The industrial age sought to marginalise even further conceptualisations of the sacredness of female sexuality. In their work *The Myth of the Goddess*, Baring and Cashford (1991) argue that former civilisations recognised a sacredness in the duality of male and female, but it can be argued that the industrialisation of the nineteenth century reinforced Judaeo-Christian traditions of denying sacredness within the sexual act and the union of male and female. It did this, argues Foucault, by the adoption of a scientific de-aggregation of the whole by concentration upon the particular. As noted, Foucault argues that the nineteenth century offered a new discourse. He wrote:

> [T]he family [became] as an agency of control and a point of sexual saturation: it was in the 'bourgeois' or 'aristocratic' family that the sexuality of children and adolescents was first problematised, and feminine sexuality medicalised; it was the first to be alerted to the potential pathology of sex, the urgent need to keep it under close watch and to devise a rational technology of correction.
>
> ([1976] 1990: 120)

Whereas for earlier generations sex was sacred, for the Victorian period sex was, according to Foucault, a pathological problem. However, it needs to be also noted that Thane (1999) refers to the slender evidence about the Victorian belief in the 'passive' woman and notes the diaries of Beatrice Webb as a counterpoint to such a view. Nonetheless, for London in 1894, the talking point was the state of the music hall, and its role as a venue for the meeting of prostitutes (Stokes, 1994). Thus, while within the societal fragmentations of the nineteenth century, just as there emerged the concept of holidays as periods of non-work as described above, additionally there also emerged the concept of the fallen woman as a source of disease and danger. This was very specifically evidenced by the Contagious Diseases Acts of 1864, 1866 and

1869 passed in the United Kingdom. S. Edwards (1997) describes the basic premise as:

> The philosophy guiding these Acts considered prostitute women as purveyors of the diseases of syphilis and gonorrhea, and children and men who fell victim were morally blameless and in need of protection through the control of these women.
>
> (1997: 58)

This interpretation is sustained by reports of the Royal Commission into the Contagious Diseases Acts. Pateman cites one such paragraph:

> There is no comparison to be made between prostitutes and the men who consort with them. With the one sex, the offence is committed as a matter of gain; with the other, it is an irregular indulgence of a natural impulse.
>
> (1988: 264)

Such views were not restricted to the United Kingdom. Sullivan (1997) points out that in Queensland the Contagious Diseases Act of 1868 was applied to all towns (unlike Tasmania), and writes that prostitutes and female immigrants were portrayed as women of 'poor character who resorted to prostitution out of a desire to avoid what was regarded as more honest toil' (ibid.: 21). Sullivan also characterises these Australian Acts as being ones where such women were perceived as the source of sexually transmitted diseases while their clients were simply disregarded. The stigmatisation of prostitutes is one that has continued to the modern era. If prostitutes are stigmatised, what of their clients? Feminists have been very quick to highlight the fact that the male clientele of prostitutes have not been, until very recently (and still comparatively leniently in the case of Britain's kerb crawling legislation) penalised by the legal systems of various countries (see S. Edwards, 1997). For example, in 1993, in the UK there were 7,912 prosecutions in the UK against women for loitering and soliciting as against 857 against men for kerb crawling. Additionally, English law has created the class of 'common prostitute' under the 1959 Street Offences Act a classification created by the woman receiving two police warnings or cautions – warnings that do not require court appearances. The consequence of this is that a woman charged with loitering may a face a charge, which is read out in court, which specifies that, 'Mrs Smith, you being a common prostitute did loiter and solicit in The Strand on January 1st 2000 for purposes of prostitution' followed by a question as to how the defendant would plead. Hence, the allegation and classification of being a prostitute, without prior evidence being offered, are made explicit, thereby denying the usual right of innocence until proven otherwise. The English Collective of Prostitutes (ECP) add to this denial of innocence a further charge, and that is the police administering the system have themselves often been racist and sexist, thereby creating a further infringement of human rights (ECP, 1997).

The importance of such labelling and stigmatisation has long been clearly understood within the literature on deviancy. For example, Turner and Surace (1956) argued that in the case of 'Zootsuiters' in California, the hostility evoked by their behaviours became only actionable when labelled with a pejorative term so as to undermine the romantic and exciting images of being 'Mexican'. In passing it might be queried if this is but a further example of the racism implicit in Western representations of non-Occidental peoples failing to repress what are deemed as unsuitable behaviour. Rock (1973: 20) argues that the deviant role 'is given a recognised place in the social structure and those who assume it are led to expect that becoming deviant will be a fateful process'. To be deviant is to be a rule-breaker, and knowing oneself to be a rule-breaker predisposes one to certain actions, to certain roles and the acceptance of deviant personality. But Rock also recognises that the rules by which deviancy are defined are ambiguous, and often rest upon power relationships within society. It thus becomes possible to challenge these rules, the assumptions that lie behind them, and to take advantage of power shifts within society.

Thus, the assumptions of the recent past age of industrialisation are now being challenged by a prostitute voice which seeks recourse to an earlier discourse, that of the sacredness of the sexual act. This will be discussed in more detail subsequently, but it can be noted on the web pages of Annie Sprinkle, the feminist, performance artist and sex worker, and found in the pages of work expressing the prostitute voice. Veronica Vera – spokesperson of Prostitutes of New York (PONY) is quoted by Bell as saying:

> Sex is a nourishing, life-giving force and as a consequence sex work is of benefit to humanity … Sex workers are providing a very valuable service to be honoured. Sex work … is a good service, it is the best service that one individual can do for another individual …
>
> By affirming sex as a nourishing, healing tool, by being tolerant of one another's sexual needs, by affirming sex workers as practitioners of a sacred craft, we accept and affirm our own humanity, an empowering act.
>
> (1994: 108)

For a sex worker feminist writer like Susie Kruhse-MountBurton, who rejects Dworkin's position on prostitution as being too reductionist, 'Men hold most of the power in society and yet they are the sexual and emotional supplicants who depend upon women's bodies for nurture and reproduction' (1996: 14). A middle way between the feminists who argue that women are manipulated and abused by men and those who reclaim the role of 'priestess' for prostitutes is represented by the view expressed by argument that the prostitute engages in strong boundary maintenance. Chapkis (1997) provides an illustration of this by citing the words of San Francisco sex worker Carol Queen, who argues:

> We create sexual situations with very clear boundaries, for ourselves and for our clients. In fact, one of the things that people are paying us for is clear

> boundaries. It's like the person going to the massage therapist; you're paying to be touched without having to worry about intimacy, reciprocity, and long-term consequences.
>
> (quoted by Chapkis, 1997: 77)

As Chapkis (1997: 76) observes, because sexuality and emotion are stripped from their presumed relationship with nature and self, it need not be assumed that subsequent alienation is destructive. The implication is that it is, in short, a job of work – one requiring skills that not everyone can do, but no more or less a job for all that.

It has been observed that the client has not suffered the same stigmatisation as the female, but this is not to deny that the client has not been criticised. For some researchers the habitual user of the services of prostitutes is a sorry sight. Thus:

> The farther we look into this hard core of single habitual customers – those with the most prostitution experiences – the clearer the picture becomes of people who are 'on the outside.' They characteristically have difficulties holding down a job or managing their own finances, and they have a limited social network. Put bluntly: the more customer experiences a man has, the more different he is.
>
> (Høigård and Finstad 1992: 39)

Similar expressions can be found in the literature relating to sex tourists (Cohen, 1986; Seabrook (1996). Very occasionally a 'happy story' can be found. Seabrook provides us with the story of 'Tony'. Tony's wife died when he was in his mid-forties, and for fourteen years he has sustained a relationship with Nok, a woman he met in a bar in Bangkok. Seabrook reports him as saying:

> I won't take Nok to America. I have thought about it, but no. She has her family here. Her children are married – they are more accepting of me than my daughter is of Nok. To me Thailand is the place where I found some consolation for a loss I thought I could never bear.
>
> (1996: 46)

Such stories are infrequent. The realities are commonly otherwise, and the problem with each such story is that it sustains the hope on the part of women like Nok for a man like Tony. In doing so it distracts from the economic and social realities of systems that sustain the form of sex tourism found in places like Bangkok, even while it reduces those social realities to the level of the individual.

Following Marx, it has been noted that the industrialisation process created a process of alienation. However, it can be argued that another process of fragmentation is also involved. Recognising claims for holidays reinforces a separation of work and non-work time. Haine notes that: 'The slow decay of the apprenticeship system and its rituals ... during the first decades of the nineteenth century sundered the connection between work and play' (1992: 469). Industrialisation separated these two worlds, but it went further. It emasculated the latent challenge of non-work lives as an alternative life-style. It commoditised periods of consumption of leisure in places other than home, and made safe such periods by the processes of consumption associated with the forms of capitalism common from the start of the twentieth century. The period of holidays, being a consumer product, no longer represents embryonic alternative lifestyles. Rather, the worker–consumer needs to continue to work in order to afford the holiday period, and remnants of the Puritan work ethic continue as we declare that 'we have earned our holiday'. Additionally, by setting non-work periods as periods of less value, it ascribed to those who did not work, mothers and the elderly, a secondary social role. Yet, in spite of these processes, it could not entirely sweep aside the potential challenge posed by holiday periods as moments of escape. The holiday retains a powerful potential for catharsis and identity confirmation or renewal. This is evidenced by popular literature, observation, personal experience and research.

For example, the novels of David Lodge illustrate the processes of renewal of self. At the commencement of *Paradise News* (1992) a description is given of garishly dressed tourists who, with varying degrees of anxiety, check in for their flights aided by world-weary couriers. Yet, by page 180 the main character is reviewing his past life, examining his religious beliefs and by the end of the novel a reborn life that comes complete with a booklet of the Hawaiian Folk Mass and a quotation from Miguel de Unamuno's *The Tragic Sense of Life*. Love, and sex, have entered our hero's life. In *Therapy – A Novel*, the relationship between self-discovery, pilgrimage and tourism is made all the more explicit as the hero, Laurence Passmore who describes himself as 'I am fifty-eight years old, five feet nine-and-a-half inches tall and thirteen stone eight pounds in weight' (Lodge, 1995: 19) and thus 'tubby', discovers lost love from forty years previously on a tourist holiday tracing the pilgrims steps to Santiago de Compostela in Spain. Similarly, Willy Russell's heroine, Shirley Valentine, discovers courage to be herself on holiday (Marsh, 1997). This author, as a windsurfing instructor in Greece (an example of acted out fantasy?) remembers a tourist who sold his business in Scotland to return to Greece to earn a living teaching holidaymakers to sail catamarans. A nurse who was a tourist one year, returned the next as a windsurfing instructor, met a young man and was last heard of heading for the ski slopes with her new partner. In an article, 'Conversations in Majorca – the over 55s on holiday', Ryan (1995) recounts meeting a respondent who described herself as being a Shirley Valentine – being a tourist, returning home, divorcing her husband, returning to Majorca and eventually marrying, happily, a local Spaniard. Wickens (1994) attributes the label 'Shirley

Valentine' to a group of hedonistic tourists in Greece, and has one such describe herself as:

> You are here to please yourself … As far as I can, I leave my everyday life behind. When I'm in England, I'm fitting into an appointed role of somebody's wife, somebody's secretary. Here, you can relax, and rub off some of the sharp corners. You are not restricted. Greeks are very tolerant of us … If you give yourself a chance, you can find out things about yourself that you did not know before … I am less age-conscious here … I like sex but not with my husband … I come to Greece for a bit of fun.
>
> (1994: 821)

Holidays are thus the marginal periods with latent potential to change people's lives. Tourists live as liminal people caught in the between and betwixt worlds of their own homes, but not in the homes and worlds of their hosts. Sometimes their holiday homes are artificial homes, architecturally designed to encourage life by the side of the swimming pool, to gaze upon other tourists and cosseted so that daily tasks of making beds and washing dishes are no longer required. Spatially isolated from the world of home and the culture of the country within which they are located, resort complexes reside in a geographical marginality, often beside the liminality of the littoral zone – itself a place that is neither mainland nor sea.

This chapter has begun to develop a thesis of sex tourism as an interaction between two groups of liminal people, people who occupy spaces between different worlds. The tourist engages in a temporary escape from the world of work, but returns to it. The prostitute exists more permanently on the edges of society, but also engages in processes of departure and return as she, or he, resumes non-prostitute roles. For both, as will be developed more fully in later chapters, issues of self-identity arise. However, this chapter has also argued that our conceptualisation of these roles has its antecedents in the development of the processes of industrialisation ushered in by the Industrial Revolution in the latter half of the nineteenth century. While, in the tourism literature, such an analysis has become subsumed in a move towards postmodernism (e.g. Urry, 1990; Rojek and Urry, 1997; Hollinshead, 1999), and concepts of de-differentiation, it is thought important not to lose sight of the conditions of the Industrial Revolution and the powerful influences they had upon patterns of life and subsequent thinking. Durkheim's book *The Division of Labour in Society*, while written in 1893 was only translated into English in 1933 and thus today we are further removed from the period when Durkheim's analysis entered the consciousness of Anglo-Saxon scholars than Durkheim himself was distant from the processes he described. Yet these were indeed revolutionary forces and they should not be under-estimated. Carlyle, writing in 1843 described the consequences of industrialisation thus:

> So many hundred thousands sit in workhouses: and other hundred thousands have not yet got even work houses; and in thrifty Scotland itself, in Glasgow or Edinburgh City, in their dark lanes, hidden from all but the eye of God, and of rare Benevolence the minister of God, there are such scenes of woe and destitution and desolation, such as, one may hope, the Sun never saw before in the most barbarous regions where men dwelt.
>
> (Carlyle, [1843] 1965)

The age created, it has been argued in this chapter, such fragmentations of society that their echoes still persist today. The processes of industry commenced then have given rise to the multinational corporations of today, and the inequalities of wealth are still perpetuated, between and within nations. As will be described, some women are exploited due to the poverty of their families, a poverty located within global systems of wealth and power. Some women turn to prostitution to deny the poverty that might otherwise be ascribed to them. For a nineteenth-century female activist like Emma Goldman, the cause of the trade in women was:

> the merciless Moloch of capitalism that fattens on underpaid labour, thus driving thousands of women and girls into prostitution … these girls feel, 'Why waste your life working for a few shillings a week in a scullery, eighteen hours a day'.
>
> (Goldman [1917] 1970)

Many still echo this sentiment today. In their turn, workers turn to tourism as an escape, a relief from patterns of work that increasingly make demands upon time and their families, and as part of that process seek relief from prostitutes. Such relief is symbolic of both processes of exploitation and succour. The processes of the Industrial Revolution, it has been argued, created the antecedents of the modern holiday by creating a demand for, and a supply of the tours offered by Mr Thomas Cook and his European counterparts. It has created a dichotomisation of women – the Madonna/harlot which fascinated and appalled the Victorian mind. The imperialism of the period also extended this classification of womanhood into an exotic other – of eastern women alluring, sensuous and equally dangerous. Sex tourism combines all of these historic connotations which are continually made explicit and implicit every time the client asks, 'How much?'. Yet, while this chapter has been based on Turner's use of comparative symbology but selecting symbols of past and present, such an analysis can only be partial. The 'Social Dramas' described by Turner (1974) have their symbolic value because of the functions they perform. The functional and pragmatic must too be assessed, and thus the question as to why prostitution and tourism go so closely hand in hand in the modern era needs to be asked. This is discussed in the next chapter.

2 Tourism and prostitution

A symbiotic relationship

In the previous chapter sex tourism was described as an interaction between two liminal peoples, peoples separated from the mainstream of society through a process of fragmentation derived from the Industrial Revolution. It was also argued that, within the semantic component of comparative symbology, they represented a signal relationship to mainstream society by postulating alternatives which meet needs denied by that society. This analysis supports the thesis proposed by Downes and Rock (1988: 203) that 'Those who deny or defy important separations and definitions within society do more than merely break a rule. They may be thought to challenge the very legitimacy and structure of order, becoming agents or instances of chaos.' But there exists too a pragmatic relationship – the relationship of client, customer and market transaction, and the market transaction possesses the role of bringing the alternative lifestyles of tourist and sex worker within the capitalist structures that dominate western society. The question has to be asked, is there an even closer relationship between tourism and prostitution than simply having historical antecedents in social forces that gave rise to the current situation? This present situation can only continue because either a mutually advantageous social exchange takes place if we are to follow Ap (1992) and his conceptualisation of social exchange theory, or, to follow Bishop and Robinson (1998), an unequal and exploitative condition has been created.

There is little doubt that such an exploitative condition exists. Even the most superficial review of sex tourism must recognise the brutality that occurs within prostitution. For example, with the growth of prosperity in China there has come a growth in pornography, prostitution and trafficking of women. *The South China Morning Post* reported the police rescuing 10,000 women from slavery in Hunan Province (31 August 1991, cited by Jaschok and Miers, 1994: 264). *The Herald International Tribune* of 7 September 1991 reported that Vietnamese women were being lured into prostitution in China (Jaschok and Miers, 1994: 265). With the growth of prosperity in China there has come a growth in pornography, prostitution and trafficking of women. The campaign group, End Child Prostitution, Pornography and Trafficking (ECPAT), based in Bangkok, notes that 'The commercial sexual exploitation of children has paralleled the growth of tourism in many parts of the world' (ECPAT, nd: 25).

ECPAT estimate there are 60,000–100,000 children involved in the sex industry in the Philippines, that 20 per cent of Vietnam's growing commercial sex industry is comprised of children under 18 years of age, and in Phnom Penh in Cambodia 31 per cent of sex workers were between the ages of 13 to 17. Trafficking in young women is reported in Bangladesh, Burma and India. Sexual exploitation of children is part of a wider exploitation as young children are forced to work in factories, building sites and sweat shops. Muntarbhorn (1996) notes: 'There can be no more delusions – no-one can deny that the problem of children being sold for sex exists, here and now, in almost every country in the world.'

There is also emerging a literature which argues that the bars of Patpong that cater for the *farang* are not entirely exploitative, but permit economic support for rural families and generate opportunities for women that might not otherwise exist. Phillips and Dann (1998) emphasise the 'white knight' syndrome of guilt assuagement by 'white' men. Thus the 'client thinks of her as a girlfriend, and she calls him her boyfriend (*feng*) ... The legitimacy of the relationship has been established' (1998: 68). The bar girl becomes an entrepreneur:

> [she] schedules the return trips of her boyfriends so that they do not coincide with each other, and she herself has become quite the cosmopolitan traveller with a passport showing trips ranging from the United States to Switzerland. It is quite a skill to be able to play such a large part in the life of a man whom she has only known for a week.
>
> (ibid.: 69)

Yet the reality of this relationship is that it is based upon a hierarchy where the benefits are assumed by only a few, and where each new, young arrival from the fields is a rival for the monies of the *farang*. Odzer (1994), who in her own research role displays the ambiguities and paradoxes of Patpong life and the exotic spell of the 'other', and who is not unsympathetic towards those who people the *soi* of Patpong, nonetheless clearly recognises a potentially cruel hierarchy when she writes:

> A hierarchy of jobs existed on Patpong. Working in a blow job bar ... or performing in Fucking Shows was at the bottom. Next came dancing nude and performing trick shows in rip-off bars ... then dancing nude and trick shows in non-rip-off bars ... Bikini dancing in ground-floor establishments was high status, but working in evening clothes without having to dance ... was higher. A distinction existed, however, between pretty women in dresses, who were brought out often, and less ravishing, perhaps fat, older women who served as hostesses only. Hostess-only types were pitied. Attractiveness and sex appeal were major elements of Patpong prestige. ... At the top of the status hierarchy were the beauties who didn't work for a bar at all but came and went on their own time.
>
> (Odzer: 1994: 65)

The hierarchy is based not on personal worth but on the dexterity of the vagina and the possession of 'good' looks. Additionally, Bishop and Robinson (1998) argue that the economic value of the bars of Patpong only continues to perpetuate an economic system based on economic and social inequalities rather than seeking to remove those inequalities. Ryan (1999), while recognising the economic and possible enhancement of self-identity of the women who work within these Thai bars, still nonetheless asks whether this is a desirable means of achieving these ends.

Such hierarchies as described by Odzer (1994), such conditions of slavery are based on exploitation, and this must be recognised and stated. This is done later in this book when examining the conditions of sexual slavery and the trafficking in sex (Chapter 5). The research of the present author has not, on the whole, been of such conditions. Most of the author's research has been undertaken in the societies of New Zealand and Australia where licensed massage parlours exist and where generally street work is but a small segment of total prostitution. Hence the extreme conditions of exploitation as found in Eastern Europe or Asia are not found. This is not to say that exploitation does not exist within these Tasman societies. As will be discussed, at one level exploitation of women, transsexuals and gays exists where some, due to low income, feel that no other choice exists but prostitution. At another level, massage parlour management can be exploitative. In one instance known to the author a female proprietor sought to impose on the sex workers a contract which required the women to pay $200 to work at a particular parlour in New Zealand. The contract also stated that non-appearance for a shift was automatic grounds for dismissal, while, in an attempt to bamboozle and frighten, the contract also included clauses such as the women being responsible for any contravention of the Resource Management Act, an Act which in New Zealand is primarily concerned with environmental protection and had no relevancy to the situation within which the women found themselves.

Within Australia there is no common legislation, with each State or Territory being responsible for its own law in this matter. The condition in Queensland has been particularly open initially to corruption and, second, one of risk for women. The 1987 Fitzgerald Inquiry in particular found evidence of large-scale corruption by the police. Inspector Allen Bulger of the Licensing Squad was found to have received AUS$274,000 from various criminal sources, while the Licensing Squad itself earned AUS$1,787,750 of 'black money' – including payments from well-known massage parlour owners, Hector Hapeta and Vittorio Conte (Khazar, 1998). Sullivan says of the Inquiry that 'evidence presented to the Fitzgerald Inquiry during 1987–88 suggests that the heavy criminal penalties for all prostitution-related activities established in Queensland facilitated the establishment of an extensive system of police graft and corruption during the 1970s' (1997: 154). This corruption extended into the heart of the political system with a leading Queensland politician, Don Lane, also being jailed for his part in the protection of corrupt officers. However, subsequent legislation which has made it illegal to work in a brothel and does not permit support services for women working for escort agencies has exposed women to risk. For example, there have been cases of women brutalised and murdered, partly, it is argued, because it is illegal for them to have drivers handy who can act on their behalf, or for them to have help at hand by having others in the same house. Thus, the condition of women being abused within western societies must also be recognised.

The main purpose of this chapter is to further explore the relationship between tourism and primarily heterosexual prostitution, but at a functional as well as symbolic level. This will be done by first briefly reviewing some of the literature that describes the use of sexual imagery in tourism advertising. Second, an argument will be made that many of the things that actually motivate holiday taking can be applied to visiting a prostitute. Given the opportunity provided by travel, it is thus not surprising that many men are able to fulfil holiday motives of relaxation and pleasure by spending time with female company. Third, by adopting a stance that the commercial transaction between two consenting adults characterises the client–prostitute relationship, the claims that commercial sex enhances personal identity will be examined. However, even within this context the conceptualisation of the liminal is not without relevance. It has been argued that the tourist is a socially sanctioned marginal person, but the sex tourist goes one step further into liminality by entering what has been constructed as the world of the profane. It is, in Sutton-Smith's (1972) terminology, a move into anti-structure. Hence far, in Chapter 1, the concept of liminality, of people on the margins of society, has been used as a device to analyse both tourism and prostitution. However, as used by Turner, especially in his work *The Ritual Process: Structure and Anti-structure* (1969), the concept applies to moments within tribal societies. He draws our attention to initiation ceremonies whereby an individual is ritually made marginal until taken in by the community. By 1982, however, Turner, adopting Sutton-Smith's views, referred to liminoid phenomena. These he saw as being characteristic of democratic-liberal societies. He writes:

> In the so-called 'high culture' of complex societies, liminoid is not only removed from a *rite de passage* context, it is also 'individualised.' The solitary artist creates the liminoid phenomena, the collectivity experiences collective liminal symbols. This does not mean that the maker of liminoid symbols, ideas, images, etc., does so *ex nihilo*; it only means that he is privileged to make free with his social heritage in a way impossible to members of cultures in which the liminal is to a large extent the sacrosanct.
> (1982: 52)

Thus sex tourism can be called a liminoid phenomenon because:

1 It is individualised and contractual.
2 It occurs at 'natural disjunctions with the flow of natural and social processes' (Turner, 1982: 54).

3 As discussed in Chapter 1, it is co-existent with, and dependent upon, a total social process and represents its subjectivity and negativity.
4 It possesses the nature of being profane, being a reversal of roles, an antithesis of the collective, but possessing its own collective representation.
5 It is idiosyncratic, quirky and ludic. 'Their symbols are closer to the personal-psychological than to the objective-social typological pole' (Turner, 1982: 54).
6 Ultimately they cease to be eufunctional, but become a social critique exposing injustices, inefficiencies and immoralities of mainstream economic and political structures.

The moment of intercourse between client and prostitute is obviously a contractual one between individuals. The language of the massage parlour betrays this in most instances. What occurs behind the door of the bedroom is 'between the lady and the gentleman' – it is not the concern of the parlour manager (Spectrum, 1998). Of course, this may in part be due to the legal situation. One massage parlour manager in Christchurch, New Zealand explained why his notice board did not list 'all-inclusive prices' on the grounds that to do so would mean that the police would interpret his parlour as being a 'brothel' and hence illegal. Rasmussen and Kuhn (1977: 13) note that 'Another important technique used to protect the masseuse from legal constraint is the word game.' The nature of the space and culture of the red light district or the massage parlour is one that normalises the process of commercial sexuality, and, as will be argued, creates its own 'collective representation' of symbols and pragmatism. It is quirky, and it can be fun. In a paper co-written by this author (Ryan and Martin, 2001) the second author, a stripper, said of the paper, 'It's fine, but we have lost the fun, the laughs of the clubs' (Martin, personal communication). In meetings with Susie Kruhse-MountBurton, referred to in the final chapter, Susie also noted that you have to have 'a sense of fun and of the absurd for this business'. Finally, it is a central thesis of this book that as the inter-relationship between sex tourist and sex worker moves from the symbolic to the pragmatic, it thence progresses to the functional. It thereby possesses a potential as a catalyst for encapsulating or creating social change due to the way that people respond to its increasing visibility. This is evident in changing attitudes towards the sex industry. Thus, for example, in New Zealand a Private Member's Bill has been drawn up which seeks to decriminalise prostitution. This compares with the perceived lack of visibility of the sex industry for *farangs* within respectable middle-class Thai society that is discerned by Hamilton (1997) and Bishop and Robinson (1998) as one reason for the sustained existence of the industry. They argue that it is this defensive lack of recognition that explains Thai resentment about Longmans and Microsoft making references to the Thai sex industry in directories and encyclopaedias (see Hamilton, 1997: 145 and Bishop and Robinson, 1998: 204) and their unwillingness to host, in 1998, the filming of another version of *The King and I.* For Bishop and Robinson (1998: 206) this 'deadening silence', this organised forgetfulness or *asphasic* gap, is deliberate. It is described as a 'will-to ignorance'; a coping mechanism at best and a political manipulation at worst. Thus the functionality of sex worker–client relationship in the Thai case is demarcated as 'an other' within Thailand itself, and by such demarcation is 'en-bounded' as beyond discussion even while its economic benefits are accepted. However, before entirely accepting this thesis it does need to be noted that *The Bangkok Post* and *The Nation*, two significant Thai newspapers, have run articles on child prostitution and child enslavement on many occasions. In addition, anyone who accesses *The Bangkok Post* Internet pages, can easily find several stories reporting problems relating to sex tourism. Thus it is the degree of the 'killing pretense', 'this privilege of unknowing' (Bishop and Robinson, 1998: 208) that must be questioned. Given the legislation approved by the Thai Parliament on 4 September 1996 with its penalties for those who procure prostitutes under the age of 18 years while creating four remand homes for under-age prostitutes for education, counselling and training, one wonders how 'forgetful' this debate is. Nonetheless, the report of the debate in the *Bangkok Post* of 21 December 1996 (Assavanonda, 1996) is not without interest as to where blame is being laid, for example, Police Major-General Surasak Sutharom is quoted as saying that 'Sex workers who returned home with their pockets full of money were one factor attracting the younger generation to jump into such illegal business.'

However, it might be objected that the coverage of sex tourism by *The Bangkok Post* does not invalidate the argument of *asphoria* advanced by Bishop and Robinson as this paper is written in English. However, it is felt their thesis can still be questioned. For example, Jackson writes:

> In researching the history of prostitution and attitudes towards women in Thailand, Scot Barmé (personal communication) reports an almost complete absence of references to homoeroticism in Thai newspapers and popular magazines from the 1920s and 1930s. This contrasts with the voluminous press reporting of heterosexual prostitution and the changing roles of women at the time, and the existence of a thriving underground history in heterosexual pornography.
>
> (1997: 169)

Certainly a considerable Thai-sourced literature exists on problems relating to prostitution and can be easily found, particularly after the initial incidence of AIDS (for example, Myo Thet Htoon *et al.*, nd: Sawaengdee and Isarapakdee, 1991; Kanato and Rujkorankarn, 1994) and little of this literature is cited by Bishop and Robinson. They appear to reach their conclusion by an analysis of economic data which is silent on the contribution made by prostitution to the Thai economy, but given the problems associated with calculating the contributions of tourism *per se* to an economy (see the discussions on satellite accounting, for example, Boskin, 1996; Smith and Wilton, 1997; WTTC, 1997; Miller, 1998), it is hence not surprising to find that the statistical database for assessing the role of sex tourism in Thailand's economy is far from complete.

Again, the thesis adopted by Bishop and Robinson can be criticised as to the politico-economic manipulation of Thailand by western interests. For example, Aramaberri (2000) estimates that the value of sex tourism to the Thai economy is approximately 1.3 to 2.5 per cent of GNP, and at most just under 6 per cent of exports. Aramaberri concludes that the use made of the economic data by Bishop and Robinson is incomplete and selective, and the 'authors use the globalization synecdoche only to pave the way for the real purpose of the book: a moral advocacy against prostitution' (2000: 242).

This is not to say that Bishop and Robinson may be wrong, but rather simply seeks to say that their thesis requires closer examination. What the debate does illustrate is one aspect of the nature of liminoid phenomenon – it is by its nature a conceptualisation of shifting boundaries, unclear definitions, differing constructions and disjunctions in society, even when as significant as sex tourism.

Sexual imagery in tourism advertising – commodities or exploitation?

It has been argued that the first link between tourism and prostitution occurs at a symbolic level through the sexual images used by the tourism industry. Oppermann *et al.* (1998) review sexual imagery in examples of tourism promotional literature in Oppermann's book on sex tourism. While examples of sexual innuendo are found, they conclude that in some countries' advertising it exists but to a very small extent. They also specifically identify tourism promotion emanating from the Tourism Authority of Thailand and conclude that 'The majority of images presented in these brochures and booklets presented visitors and locals in realistic and modest dress and their actual less-than-perfect appearance' (Oppermann *et al.*, 1998: 28–29). Two immediate observations may be made about this work. First, it does not use the word 'gender' and thus the analysis is undermined by a failure to distinguish between gendered representations and those of a more sexual nature. Second, the research was limited to the official promotional material of tourism authorities, and thus excluded from consideration the more explicit advertising material that is aimed at sex tourists.

Marshment (1997) also reviews the use of sexual imagery in the advertising of mainstream tourism products, namely the products of British package holiday companies. She argues that while there are images of the woman in a swimsuit, such a figure 'signifies the pleasures of idleness and luxury that certain representations of the beach holiday offer. There are no comparable images of men … she is not part of the package holiday, but a signifier of it' (1997: 20). Marshment argues that at face value gender is not a factor in the selling of holidays, but in these cases of package holidays there are gender constructions of definitions of the marketplace. The market is the nuclear family, they are heterosexual, with ideals of companiate marriage and are associated with a range of definitions of ordinariness. For Marshment, however, there is evidence of just how gendered are the representations of our culture. She argues that the embodiment of pleasure is the female body, not the male, and that is why it is the female body that is used to represent the pleasure of the beach holiday. Even more specifically, it is the young, slim female body, while the exoticism of the 'other' is tamed by the use as a signifier of the young, slim, and *demure* female figure.

Considering that these images are derived from mainstream holidays, arguably these results are not surprising. If one is looking at the holiday images of British package holidays which are aimed at the mass market, then the use of homosexual imagery is going to be primarily absent. However, if one resorts to the media aimed at the gay market, magazines such as *Boyz*, *Cafe* and *Attitude*, then one does find pleasure being signified by young, slim *male* figures. Equally the messages of sex tourism are generally far more explicit. How explicit these images will be are partly constrained by the media being used. Press advertising for Auckland's massage parlours offers a range of images. Some refer to gentlemen's clubs with nothing more than a picture of the premises, but many feature pictures of the women with short descriptions. These emphasis bubbly personalities, fun women to be with – thus 'Suzanne is Asian and aims to please and tease', while 'Terri is not shy and nor should you be'. Some indicate specialities or offer 'toys' (advertisement in *New Zealand Truth and TV Extra*, 31 July 1998). While the photographs are of women wearing little and being topless, these are not passive women. Indeed, domination services are offered. The images contrast with those offered by the videos of Dexter-Horn productions. In the case of *The Erotic Women of Thailand* (Dexter-Horn Productions, 1997) images are presented of seemingly embarrassed young women having showers with a voice-over which emphasises that these are not models but real women that you can meet. Towards the end of the video is an explicit appeal to men who are disillusioned with 'demanding, liberated North American women'. Thai women are 'real' women because they are submissive. O'Connell Davidson (1995) traces similar attitudes of British sex tourists in Thailand. This implies that the market for these 'exotic–erotic' trips are men unable to cope with the self-confident women of the West, and there is evidence for such a thesis. Seabrook describes the situation thus:

> Westerners think that sex is the ultimate authentic human experience. The young women and men of Bangkok know better. Because the farangs are so lonely, because they are such isolated individuals, they imagine that it is through sex that human beings come closest to one another. They think they see the profoundest communication in what is the loneliest experience in the world. They think like this because they are so far away from each other, they have to reach across the empty spaces of their separateness before they touch another human being.
>
> (1996: 114–115)

What strikes this author about a tape like that of *The Erotic Women of Thailand* is just how little fun the whole process is shown to be. The shots taken in the clubs have, judging from the voice-over, an illicit thrill not because of the scenes they show, but because filming is not permitted and thus it is an 'adventure' to take such shots. The result is one of grainy shots of poorly lit rooms showing bored women in their underwear. The images are not sexy or erotic and if pornographic are thus because of the allusions being made. That they are thought to be erotic says much about the state of mind of the viewer. (Of course, such a judgement is itself subjective, and the issue of researcher subjectivity in such a subject was raised in the Preface by the authors.)

If we are to look at sex images in relation to sex tourism, then the search for imagery has to go beyond the sources of imagery analysed by Oppermann *et al.* (1998). However, as noted, the nature of the image presented is modified by the media in which it is presented. The imagery presented by small adverts in easily accessible newspapers differ from those in men's magazines, or in the magazines aimed at lesbian, transsexual and gay readerships, all of which are more explicitly sexual. In heterosexual imagery 'close-up action' is often promised. Some women advertise 'back entry' or an experience of 'showers'. The photographs are of topless women; some offer 'double action' or 'voyeur sessions'. Touts in the *soi* of Patpong offer leaflets describing fucking shows (Odzer, 1994). Gay readers are offered just as great a range of services as their heterosexual counterparts. Detailed guides exist, for example *The Best Gay Guide to Amsterdam and Benelux* (1992). If this is thought to be offensively explicit it is of interest to note that while writing this chapter, left in a staff room the author found a copy of *Cleo* for August 1998. Articles in this women's magazine included 'Your vagina – what's normal'? and the latest in pubic hair designs. On page 36 there is an article with the headline 'Open wide, come inside – it's Love School' – an article devoted to obtaining a good orgasm and which is illustrated with a blow-up male doll complete with a large penis, clitoris jewellery and a vibrator. For the virtual sex tourist of cyberspace, as is described by Kohm and Selwood (1998), there exist a full range of sites which provide information, pictures and reader's reviews. Explicit sexual imagery is easily found. What is of interest is the interpretation of such imagery. For many the images are simply pornographic. But if one uses the dictionary definition of pornography as being the explicit description or exhibition of sexual activity in literature, films, etc. intended to distinguish erotic from other, more aesthetic, representations, then several problems quickly emerge. First, does such an intent exist, second, how does one judge the success of such intent, third, is there need to show the outcome of the intent to stimulate? Sullivan (1997) demonstrates the problems involved in attempts to legislate such imagery. She describes the anti-pornographic view by excerpts from Parliamentary debate: thus

> They came to believe that pornography is a violation of the rights of women. They found that the majority of all this material is based on treat-

> ing women not as persons but as things, and on degrading and humiliating women.
>
> (SAPD 1977–78: 1292–1293, cited by Sullivan, 1997: 169)

Yet, as evidenced by the example of *Cleo* for August 1998, it is possible to publish an article for the general readership of women which ostensibly would have at least a purpose of titillation. Thus the article goes on to note that the school's objectives include providing 'an arena in which single people can meet each other. The school is always booked solid around Valentine's day' (Broadbent, 1998: 39). Sexual freedom is being increasingly constructed around females being able to enjoy their own sexuality. With reference to *Cosmopolitan* magazine, Janet Lee has remarked:

> *Cosmo* exhorts the female reader to construct herself through self-discipline, and the reward for this is physical pleasure – or, more specifically, sexual pleasure ... The 'new woman' *Cosmo* version is the sexy woman. Sex equals not only fun, but independence and success. And *Cosmo* claims to have the knowledge that will tell you how to have it all – sex, success and liberation.
>
> (1988: 169)

Hence, the images of sex available to females that were once comparatively difficult to access in the mass media are now very accessible to women. A further examination of the sexual imagery associated with sex tourism reflects a further complication to the thesis of marginalisation that has been advanced previously. That thesis has been based upon a conceptualisation of fragmentation, of differentiation. The language of sexual imagery in sex tourism as derived from the actual advertising of commercial sexual services both reinforces and counter-balances that thesis. It reinforces the thesis because at one level it represents a change in the nature of the tourist's liminality. The images once constrained to pornographic magazines are now to be found in women's magazines, and the by-product of this is to drive the commercial sex industry to ever more explicit images until there is little more left to exploit. Dragu and Harrison (1989), writing from the perspective of the former being an ex-stripper and the latter a writer of pornographic stories, observe that in the nineteenth century, there was a time when even the mention of women's legs was considered to be shocking and indecent. They note:

> We have always focussed our suppression of sex on its purely physical aspects, and so it is natural that we also pursue sexual revelation at the physical level ... This [is] a perfectly valid process, and one that has served us well for a long time, but we seem to be reaching the end of the line ... Today, even a good long look at a woman's inner labia is loosing its charge.
>
> (1989: 59–60)

They argue that the belief of sexual thrill lies in the realm of what is forbidden, and as that boundary moves, so sexual entertainment may be moving into decadence. A by-product of this is that such entertainment increasingly fails to satisfy and the concentration on the purely physical denies the spiritual and emotional – they liken it to fast food, we keep consuming but it never satisfies. Furthermore, what is emerging is the growth of the female as the consumer – they too emulate their male companions as searchers for sexual entertainment. However much both males and females learn that masturbation can be satisfying, love-making is better with a partner. The past boundaries and roles between sex tourist, commerciality, gender of sex tourist and provider of sexual service, consumer and provider of sexual entertainment, are being eroded.

It has been argued that the tourist is a marginal person, but that the marginality is sanctioned by society. But by becoming a sex tourist, by responding to the imagery of sexual service, that tourist might be said to be transgressing from the licit to the illicit. The tourist now begins to share the possibility of condemnation. Ryan and Kinder (1996a) stress the importance of concealment of the client. In an analysis of Auckland's red light district they write:

> From the viewpoint of the tourist seeking a passive or active sexual entertainment ... Fort Street represents a safe 'crimogenic' place. Indeed, in many senses the term [crimogenic] is perhaps inappropriate for Fort Street. It offers a high degree of physical safety.
>
> (1996a: 33)

They go on to comment, using the concepts of 'Routine Activities' (Cohen and Felson, 1979) and Felson's (1986) concept of 'Guardians', that the tourist seeks anonymity and in doing so will 'adhere to a set of rules, for to do otherwise is to risk the concealment necessary' (Ryan and Kinder, 1996a: 34). Thus, by transgression of the boundary between the sanctioned marginality and that not so sanctioned, the tourist now requires the sex industry to provide safety and confidentiality. The reality of this provision thus permits enjoyment of a leisure pursuit – the cocoon of confidentiality permits a legitimisation of the act of patronage by those within the demarcated zone, which is both spatial and social. Hart, in her description of a Spanish *barrio* or neighbourhood where prostitutes worked described it in this way.

> Within the barrio, sex as leisure was not an unambiguously illicit or illegitimate pursuit. Many clients did voice misgivings about being there (on a periphery), and often went so far as to describe their presence in the barrio as a 'vice'. However, they were able to enjoy this 'vice' in an atmosphere in which this was accepted as a leisure pursuit, albeit one that was considered to be rather different to others.
>
> (1995: 219)

Equally, this conceptualisation of a safe space can operate at an emotional level. Thus one sex worker has written

> Sometimes, I start to wonder – maybe I do function better institutionalized – closed off from the outside world, in my own private fantasy land where I can do and say as I choose, and speak from my heart and touch and feel my humanness and not be stopped by mind barriers and should's and don'ts and all those other cultural conditionings, where the mind and logic take over for the heart and compassion.
>
> (Sisters of the Heart, 1997)

In research one client touched on this when he said: 'The parlour is where I relax, no-one knows I'm there, it is my time, my hour out of the hassles, I'm not anyone but me having a nice time with nice company.' This pursuit of sex as a leisure activity, this placing of prostitute and tourists within the same domain of the space of prostitution has many implications that lead us to conceptualisations of de-differentiation.

Competing definitions of the commercial sexual space exist. Prior to advancing a thesis of the de-differentiation offered by the locus of client–sex worker interaction and the imagery and pragmatism associated with it, it is necessary to state the arguments and views of those who will have none of what they perceive as self-deluding semantics. Kathleen Barry has played a leading role in the Coalition Against Trafficking in Women (CATW). The Coalition wishes to further enforce the 1949 Convention of the United Nations of 2 December, entitled 'Suppression of the Traffic in Persons and of the Exploitation of the Prostitution of Others' by creating a freedom from all sexual exploitation. The Coalition's web page states its aims as being a world 'where prostitution and sex trafficking do not exist, where women are free and equal in dignity and rights, where no woman is sexually exploited' (http://www.uri.edu/artsci/wms/hughes/catw/catw.htm October, 1998). Barry herself in her book, *Prostitution of Sexuality*, makes her views quite clear. Thus, for example:

> A lover, husband, or boyfriend who promotes the sexual exploitation and commodification of women is a pimp, and together, pimping and procuring are amongst the most ruthless practices of male power and sexual dominance. These practices go far beyond the merchandising of woman's bodies for the market that demands them; they crystallize misogyny in acts of male hatred of femaleness as rendered into a commodity for whom the marketer and the purchaser have contempt.
>
> (Barry, 1995: 199)

For an activist like Marcovich, a founder of the Movement for the Abolition of Prostitution and Pornography, prostitution is akin to slavery. She describes it as a market system based on a society which legimitises this market, on the actions of pimps, money and

> the clients, always men, who buy the sex act, pieces of body: vagina, anus, breast, mouth and hands of women …
>
> the prostitutes, women who are victims of this market, who, to support the violence of being denied as human, of being penetrated and tortured 1, 10, 100 times a day by men for whom they have no desire, split their spirit and their body in two, become addicts to drugs or alcohol, when becoming ill much later with anal or vaginal cancer.
>
> (Marcovich, 1998)

From these perspectives the issue is simply one of male exploitation of power. The sexual imagery of the sex industry is not about questions of identity, but solely about the exploitation of the female body for pecuniary gain by men.

The almost diametrically opposite viewpoint is stated by Murray. She argues that feminist arguments of the type postulated above are based on a false premise for there is nothing inherently wrong in the commodification of sex. Further, she argues, academia has given credence to such arguments of commodification and thus legitimised them. But Murray also notes 'The academy has progressed from women's studies to gender to sexuality, getting closer to the cunt of the matter while continuing to marginalise class, race and alternative subject-voices' (1998a: 70). For Murray the voices and images of commercial sex are not yet diffuse enough. Nonetheless she discerns that at last:

> Dyke whores are no longer double deviants, in some parts of the West at least. After being invisibilised by some feminisms, dyke whores have come out in a babble of trendy deviances, though the working-class junkie whores are still invisibilised. There are new games to play, where the referee is not the only one with a whistle.
>
> (ibid.: 74)

The imagery of commercial sex and sex tourism is becoming more diverse and the *flâneur* and voyeur is no longer simply male or heterosexual (Craik, 1997; Hughes, 1997; Munt, 1998; Pritchard *et al.*, 1998). The differentiation of whore/madonna of the nineteenth century is becoming the subject of a post-modernist de-differentiation where madonna reads *Cleo*, can respond to the advertisements that feature the services of 'Tony' or 'Steve', admire the Chippendales and other male-based sex acts aimed at women, and where the whore is butch, or gay, or black and definitely proud.

From the position of those feminist groups who work with sex workers, the views of CATW seem extreme. Lisa Hofman, Director of the Dutch Foundation Against Trafficking in Women, is cited by Chapkis as saying:

> We were shocked when we went to an anti-trafficking conference in New York in 1988 and discovered how out of touch with working women the U.S. Coalition ... seemed to be. I think it's very significant that that particular group only works with women who have already left prostitution.
>
> (Chapkis, 1997: 62)

Hence, in conclusion of this section, there exist many more explicit sexual images associated with sex tourism than those examined by Oppermann and his co-authors in his book. Essentially, the chapter by Oppermann *et al.* (1998), in the view of this author, is not about sex tourism at all and thus perhaps is misplaced. To find the images related to sex tourism it is necessary to examine the images of the sex tourism industry, and as might be expected, these images are diverse. First, they are often 'explicit' by the norms of 'normal' society – but as has been pointed out by reference to a popular women's magazine, these norms themselves cover a wide range of acceptabilities. Second, the images of the commercial sex industry are not solely related to heterosexual sex or straight sex. Third, as illustrated by the statements by Murray, those portrayed in the images do not feel necessarily degraded by the images. As will be discussed below, they can feel empowered. Images imply recognition, and images are picked up, used and, yes sometimes abused, by the media operating in mainstream society. However, sex workers have known through discrimination that they have never entirely controlled the images of their industry. The paradox is that as images become adopted, distilled, disseminated, they assume their own lives which become divorced from that which was illustrated, but in doing so enter a public consciousness which permits the prostitute voice to emerge from behind the image. As Kempadoo (1999: 27) writes of sex work 'being simultaneously a form of domination and exploitation as well as a place that enables assertions of freedom in the context of oppressive racialized economic orders'. As will be argued, there exist links between the 'macro-context' of the social understanding of sex tourism, and the 'micro-level' of its meaning for individual actors. Foucault notes in *The Carceral* that 'the modelling of the body produces a knowledge of the individual, the apprenticeship of the techniques induces modes of behaviour and the acquisition of skills is inextricably linked with the establishment of power relations' ([1977] 1996: 422). It is now to the 'apprenticeship' and the functional relationship that we turn.

Motivations for holidays and/with commercial sex

Some of this section will draw upon past research published by the author, either alone or with co-authors (Ryan and Kinder, 1996a, 1996b; Ryan *et al.* 1998; Ryan and Martin, 2001; Ryan, 1999) and from previously unpublished notes relating to that research. It will also cite other work that generally confirms these findings, but it is again important to establish a caveat on the location of that research. It is primarily based in New Zealand and Australia with some insights from the UK and thus is not immediately applicable to situations outside of those countries. Some generalisation is possible, but care must be taken when seeking to apply the findings to other regions of the world. As Hall will show in Chapter 6, the position in South-east Asia and Eastern Europe may mean that some of these findings will not be applicable to those parts of the world, or to some aspects of sex tourism such as that relating to paedophilia.

Many researchers have examined the motivation for tourist behaviour. Crompton (1979) notes that travel consumers are not motivated by the specific qualities of the destination and its attractions, but rather by the broad suitability of the destination to fulfil their particular psychological needs. He proposes that 'instead of distance, culture and climate being used to classify destinations, one can envisage clusters of vacation centres which are predominantly self-exploration, or social interaction or indeed sexual arousal' (ibid.: 409). In this sense the term 'sexual arousal' is being used literally and not as a metaphor, and it may be argued that parts of Thailand, Amsterdam or any red light area meets this definition. Mathieson and Wall (1982) refer to the physical motivational category which includes motivations such as refreshment of body and mind and pleasure – fun, excitement, romance and entertainment. It is not hard to extend these motivations to the client visiting a prostitute. Yiannakis and Gibson (1992) developed a typology of holidaymaker clusters based upon three different continuums. The first is the desire for a vacation that is either highly structured or has little structure. The second axis indicated whether the tourist preferred their tourist destination to be a stimulating or tranquil environment. The final axis was a continuum to determine whether the tourist aspired to visiting a strange or familiar environment. Hence the tourist seeking contact with prostitutes might choose a vacation with a structure that permits flexibility, seeking a stimulating environment, but also one with familiarity. They describe the 'action seeker' as 'Mostly interested in partying, going to night clubs and meeting the opposite sex for uncomplicated romantic experiences' (Yiannakis and Gibson, 1992: 33). This type of tourist has been previously described in Chapter 1 with reference to the work of Wickens (1994). Wickens also returns to the same subject in her later work of 1997. Describing a category of tourist which she terms 'Raver' she reports one as saying:

> Without the emotional baggage of love, I enjoyed the lust in the brief sexual liaison I had with Kosta. I met him in a bar … He sent a drink across to me … and ended in bed. No, I didn't experience any emotional pain when we parted a few days later!
>
> (1997: 157)

This example is telling as it offers evidence of the de-differentiation modern tourism offers to its clientele. Compare this statement with that of a former prostitute, Maryann, as recorded by Chapkis:

> I think that the assumption that being a prostitute ruins a woman's experience of sex is part of that [i.e. the view that sex can only mean one thing to women]. Men need to think that women can't have sex without intimacy, and that if they do that it's bad for them. Like a woman only has sex with a man because he and he alone has something she can't live without. In fact, an important part of prostitution for me was realizing that sex didn't have to be about intimacy. There is great power in the realization that you are, in fact, in control.
>
> (Chapkis, 1997: 84)

In much of the debate on sex tourism it seems as if the argument has been hijacked by a feminist rhetoric within which the client is the male and the prostitute female, and the relationship is heterosexual. It also implies that the prostitute is the victim. These scenarios are incomplete. For example, Albuquerque (1998: 109) describes how Barbadian beach boys size up potential female clients and 'a relatively wealthy, attractive, thirty-something French Canadian tops most lists'. At another level the issue of degradation is sometimes perceived by the client. Thus Kruhse-MountBurton notes that:

> prostitution in the Australian context is often appraised by clients as deficient, in that prostitutes are criticised for being emotionally and sexually cold and for making little effort to please, or to disguise the commercial nature of the interaction.
>
> (1995: 193)

Additionally, not only may the client be female, or gay, but the prostitute a lesbian while meeting the needs of heterosexual male clients (Murray, 1998a). It seems as if in the nineteenth century the existing hegemony stated that for a women to have sex outside of marriage was itself bad for the woman and her soul. The nineteenth century is also full of examples of women for whom to have sex without love was also equally harmful (for example, Anna Karenina or Madame Bovary). Wickens's example is of a woman who wants and enjoys sex without a wish for long-term relationships. One of the modern myths is that it is harmful for a woman to have sex without desire; that, in short, it harms a woman who commodifies her sexuality. However, from an increasing number of interviews with sex workers the evidence mounts that this is simply not the case. For example, in interviews with New Zealand prostitutes, Jo said, 'I did it because I wanted to, it was as simple as that. And when I didn't want to do it, I stopped.' Terri said, 'If I didn't want to do this, I wouldn't be here – it's a job of work.' Amber, an exotic dancer of ten years' experience said, 'It's really very simple at one level – it's just a job of work, it pays the bills.' This is not to say that the work cannot be fun or enjoyable. Sharon notes that on one occasion she had such fun with a 'real Italian Romeo type' that she let him off free of charge, but quickly adds that this is not a normal circumstance. The old question, do prostitutes enjoy it, can elicit many answers, but the stereotypical responses of 'no they don't' denies the concept that sex without love can be fun, and that as a job of work, there are good days and bad – and on a good day it may be possible to have a good time with a client. Jo Doezema, an Amsterdam-based prostitute says:

> So there are parts of my life I don't want to share at work. So what? Do I have to give all of myself and not hold anything back in order to legitimately be able to say that I like my work?
>
> (cited in Chapkis, 1997: 122)

She goes on to make the point that if prostitutes say they enjoy their work, their view is dismissed as one whereby the prostitute does not realise she is being destroyed. Burrell (1997) reports that at a conference in the UK, many prostitutes were angered and dismayed at what they felt were simplistic feminist arguments portraying them as victims. It is, quite simply, a role that is rejected time and time again by many working women.

The tourist may seek fun while on holiday as a source of relaxation. Is the person who waits at the table stigmatised for providing a fun ambience while serving them? Generally the answer is no. Can it be doubted that the waiter might not, on occasions, actually be having fun while serving a group of cheerful holidaymakers who make his or her task that much easier? Again, the answer is no. Why, then, the difficulty in believing that those sex workers able to work in an environment which generally respects them as a person and which provides her or him with control, might not also have some fun, or be professional in the service being provided? To deny the concept of sex work as being a professional service, as having like other jobs its good or bad days, is to continue to marginalise and stigmatise. When this author asked a member of the New Zealand Prostitutes Collective why she persisted in her role, the response was to bring about a 'normalisation' of sex work. Yet perhaps marginalisation is 'safer' for many people, because, as already noted, not to stigmatise means a need to recognise that non-prostitute women may also wish to either have male sex workers, or to seek male patrons for purposes of their own. That, in short, the bourgeois repression of sex in Western society may not have advantages, but in fact may psychologically cripple (e.g. see Stoler, 1979). Wickens's 'Ravers' may not be paid in money, but they are paid in terms of having the good time they desire.

Around prostitution are created many fantasies. Some are sexual, some are cultural (how often has the whore with a heart of gold been portrayed in Hollywood films? Even Rhett Butler in Margaret Mitchell's novel *Gone with the Wind* had the support of his good whore). Tourism too is a time of fantasy. Disneyland peddles its fantasies and myths of American culture – what Hollinshead (1997) has termed its 'Distory'. Tourists arrive at Disney and other locations to be immersed in myth, to play out roles. Today photographers offer services whereby they will create photographs of us all as models or figures of fantasy. What previously had been a private fantasy becomes projected onto our film or in a set – how close to reality are the worlds portrayed by the film, *West World*? Certainly, Ryan and Kinder (1996a) found evidence of clients seeking fantasy. They cite the example of 'Tony', who notes that 'it is more exciting, more uninhibited and there is no holding back. I enjoy the whole experience, the sex, the fantasies, being able to "talk dirty" and the fact that there are no demands on me.' Winter (1976: 10) comments: 'Some clients thrive on the ability to engage an anonymous prostitute for sexual relations: to them the whole experience is a novel sexual adventure filled with surprises and fantasies.' However, for the sex worker, there may be little such novelty or excitement.

Like holidays, a visit to a prostitute can meet relaxation, social and friendship needs. Evidence of this emerged from conversations with clients reported by Ryan and Kinder (1996a). 'Charlie' stated that he sees ladies to 'get a little bit of happiness'. 'Fred' commented : 'I just usually want a cuddle and some company. It gives me friendship and some social activity.' 'Nick', a business traveller who often visits prostitutes stated that he used 'high class' escorts who are usually well educated, that he had a need to interact with them on an intellectual level – that 'it is more important to be with the girl and have intelligent, cheerful conversation, than just to have sex'. Additionally, tourists visit different places in order to see and do new things – they search for novelty. So too do those who visit prostitutes. 'Sam' met escorts 'because of a search for variety'. Michael 'feels the need to do something different'. Peter, who frequently uses prostitutes when on business trips, commented: 'yes, my sexual [activity] is different than with my normal partner ... [it] involves excitement at the unknown'. Again, 'Basil' stated: 'Trying to get more of a cover of what different people are like – mainly for the variety.'

Holidays have also been noted by Crompton (1979) as opportunities for regression into childhood. The tourist visiting a prostitute may also be engaged in another form of regression. Sheehy (1971) has made the observation that, for a man, prostitution represents an opportunity of 'buying the nostalgic illusion that things are how they were when he was a boy'. Further, if holidays present opportunities for the many to enjoy, however limited the time, the life-style of the wealthy, so too sex tourism, particularly in Asia, permits an exhilaration usually open only to the wealthy, that is of having access to many women of youth and beauty. It can also be regarded as an acting out of the fantasy of being powerful. Lindi St Clair is quoted by Thomson (1996) as saying, 'Let's look at half these politicians. Let's look at half the royals. Who would give them a second glance if they weren't rich and famous? Power makes an ugly man attractive.' And every man, ugly or not, who walks into the massage parlour has the power of paying the fee. Within this act lies the repugnace felt by those feminist commentators like Barry who see prostitution as being the act of purchase of female body parts, but the parlour does offer its culture and its norms, and within this environment the sex worker does exercise considerable elements of power as will be discussed below.

The 'holiday romance' that can be found in literature and fact (although not often researched) has several features. One is that it is understood by both parties to be temporary – that its boundaries are those of the holiday place and period. There were examples of respondents in the Ryan and Kinder research, as with Wickens's example cited above, who simply wanted sex without emotional involvement. Indeed, parallels could be drawn between the actions of some younger respondents and the type of tourist featured in media portrayal of 18–30-year-old holidaymakers (Benny Dorm, 1988). 'Ted', aged 25, was one such example. He went to a prostitute the first time two years ago when 'out with the boys and we were drinking'. He had no time for emotional involvement with a girl; 'I don't want to have to go through all that bullshit.' However, this attitude was not found to be common, and was primarily shown by younger clients (Ryan and Kinder, 1996b). Both Seabrook and O'Connell Davidson describe the type. O'Connell Davidson (1995) divides the clientele into three, 'Macho Men', 'Mr Averages' and 'Cosmopolitan males'. 'Macho Man' looks to have sex with as many women as possible. Seabrook (1996: 36) describes the short-term tourists to Patpong as 'extremely insensitive' and having 'little imaginative understanding of the people whose lives touch theirs' … "A shower, a shag and a shit, the three biggest pleasures in life," said one man with his mates.' However, Seabrook goes on to write that behind the bluster, the attempts to assuage the guilt in drinking, even these men at an individual level 'become more thoughtful, and are interested in the lives of the women who service them; but they feel that by giving a generous tip, "treating them decently", they have acquitted themselves of any debt to the women' (ibid.: 36). There is a significant literature relating to sex tourists who want friendship with young women. Kruhse-MountBurton (1995: 194) cites one interviewee as stating: 'It's true there is no chance of rejection. But now maybe I'm a bit idealistic in the sense that I think, wouldn't it be nice if during that day, and that encounter, that there developed a genuine friendliness.'

Cohen (1986) reproduces letters sent by former clients to the Thai women with whom they have shared time. In the Ryan and Kinder research similar motives were also stated by respondents. Thus one man said: 'It's like meeting a girl friend for the first time … the affection may be purchased, but I am continually pleasantly surprised by just how nice the women are, and that I very much appreciate.' The difference in attitudes, between seeking or rejecting emotional support is possibly demonstrated by the language used by respondents in Ryan and Kinder's work (1996a). Those who share 'Ted's' views seemed to have a greater tendency to '*use* a prostitute'; others would '*visit* a prostitute'. Such language difference can be held to be significant, and Ryan and Kinder offer a simple content analysis by frequency counts of use of various expressions. However, such a simple analysis remains but that, simple, as it does not take into account the point made by Seabrook (1996) that within other contexts most males seem to incline to at least some reflexivity as to their role as clients.

Another reason advanced for holiday taking is that of ego and status enhancement (Crompton, 1979; Mathieson and Wall, 1982). Such motivation can also be discerned, albeit perhaps indirectly, with sex tourism. O'Connell Davidson (1998) provides evidence from her interviews with sex tourists in Cuba and Thailand which provide support for this motivation being present. Thus she cites one white British sex tourist in Cuba as saying:

> It's funny, but in England, the girls I fancy don't fancy me and the ones that do fancy me, I don't fancy. They tend to be sort of fatter and older, you know, thirty-five, but their faces, they look forty. But in Cuba, really beautiful girls fancy me. They're all over me. They treat me like a star. My girlfriend's jet black, she's beautiful.
>
> (O'Connell Davidson, 1998: 169)

For O'Connell Davidson a complex set of identities arise in such cases. The men value a certain form of female identity, they are able to possess that through economic power, they attribute value to the person 'possessed', and thereby revalue themselves. An implicit theme within this analysis is the psychological immaturity of men who are unable to see past the physical. Yet what is paradoxical in such situations is that within the situation men will often rationalise the relationship in terms of the senses of friendship, companionship and concern they will feel for the woman concerned. It raises the question whether the re-evaluation of self on the part of males actually permits the better side of their character to emerge. Such a thesis would support the priestess function described by writers like Bell (1994), Jordan (1991), McLeod (1982), and Delacoste and Alexander (1987). Unfortunately, to sustain this thesis would require evidence of a behaviour change when men return home from locations like Thailand and Cuba, and what evidence that does exist seems to support Cohen's (1986) contention that, on the contrary, men simply seek to return to the land of their beautiful women. Cohen begins his paper on the correspondence between *farangs* and Thai girls with a quote from Grey, namely:

> Excessive love for the exotic can destroy the white European in the Orient. Many men think they go away from here with their souls intact – but they find in their own countries they've been profoundly changed by their experiences without knowing it. They become outcasts among their own people because everything at home seems insipid in comparison with the East. Then they're lured back by the siren call of what has already ruined them.
>
> (Grey 1983: 254)

However, it has to be observed that such behaviour is not unique to sex tourism. Ryan (1997) relates stories of how the cathartic experience of tourism lures people back to the holiday destination. As stated in Chapter 1, and citing from his own experience as a windsurf instructor in a holiday destination, he identifies the nurse who returned as an instructor, the personnel manager for a large British retail chain who did the same, the man who sold his business in order to teach people to sail catamarans, the couple who started a cycling company in order to live in France. The liminal possesses the ability to create a permanent marginal status wherein the ritual of the liminal provides psychological support.

Another motivation that emerged from the New Zealand research (albeit associated with concealment needs) was role of prostitutes in family bonding – a need recognised in the tourism literature as one met by holidaying (Crompton, 1979). For example, 'Roger' is happy in his marriage, but feels a need to visit prostitutes for a sexual relief. He would never want his wife to find out as it would ruin his marriage and trust between him and his wife. For almost all the men involved, concealment and discretion were of paramount importance, often because of a need to sustain a marriage.

In this listing of motivations, there are leitmotifs, namely, the desire for sex and concealment. The desire for sex is, as already noted, not unknown to observers of tourist behaviour. Nor, but from a different perspective, is the desire for concealment. Pizam and Mansfeld (1996) illustrate the link between crime and tourism, and the fact that tourist locations can offer concealment for criminal action. In the context of the tourist and the prostitute, the tourist location offers the concealment of anonymity which reduces ties of responsibility. However, what emerges from the discussions with clients and prostitutes is, for some, a need to conceal actions from cared for others. Clients wish to conceal their actions from their partners, and for many prostitutes, there is a wish to conceal their actions from parents and/or children. As already argued, such a need arises from a societal viewpoint that while recognising the existence of sex work, many still prefer to marginalise it to the 'safe' location of the massage parlour or red light district in order to avoid questions relating to sexual identities and relationships within mainstream society.

To summarise this section thus far, it may be said that all the motivations that exist for holiday taking – relaxation, fantastical escape, family bonding, adventure, doing something different – all of these motivations equally apply to visiting a sex worker. The other reason for visiting a sex worker is the search for sex, and it has been often argued that the atmosphere of holidaying is signified by female figures (Marshment, 1997). While this chapter has tended to agree with Marshment's contention that mass package holidays may be gendered in imagery other than overtly sexual in appeal, it needs to be recognised that within popular literature as well as the academic, the fourth 's' of the holiday after sun, sea and sand, is indeed 'sex'. Baillie notes that:

> Tourism promotion in magazines and newspapers promises would-be vacationers more than sun, sea and sand; they are also offered the fourth 's' – sex. Resorts are advertised under the labels of 'hedonism', 'ecstacism', and 'edenism' ... One of the most successful advertising campaigns actually

> failed to mention the location of the resort: the selling of the holiday experience itself and not the destination was the important factor.
>
> (1980: 19–20)

However, it is not necessary to look for relationships between tourism and the sex industry solely in terms of the hedonistic advertising adopted by the industry. Simply put, both holidaying and visiting a prostitute or visiting a strip club may be regarded as forms of leisure activity. O'Connell Davidson (1998) reviews that form of sex work which she describes as ritual reinscriptions where young men are introduced to sexual knowledge. She writes that:

> Although even the most reluctant client may go on to become sexually excited by the process of selecting and then exploiting a prostitute, it is important to note that participation in the form of prostitute use [of this type] ... is not generally *motivated* by any particular sexual desire or desire on the part of individual participants. Nor does it involve sexual abandon.
>
> (1998: 166)

A further factor enforces the linkage between tourism and sex work. This is opportunity. Ryan and Kinder (1996a) seek to explain this by reference to two diagrams reproduced as Figures 2.1 and 2.2. The first is derived from Tonry and Morris (1985) whose work was based on studies of criminality and deviance, especially with reference to 'opportunistic' criminality. The argument is advanced that the readiness to undertake what was termed an 'unworthy act' arises from an interplay between background, needs, past learning and an evaluation of whether it is possible to both commit the act and escape undetected. It is worth highlighting that opportunity alone is not a sufficient variable. There exist a series of salient factors that shape a reaction to the 'chance event'. In the case of sex tourism it may be argued that the opportunity to visit a prostitute arises within a context of hedonism, relaxation and escape. However, this context and opportunity are again not sufficient unless the need exists to an extent where a readiness to act is formed. That readiness may be in a context of anonymity, but as Seabrook (1996), O'Connell Davidson (1998) and Ryan (1999) make clear, in some cases that anonymity is set aside in the pursuance of male rites of mateship or togetherness. With reference to Figure 2.1, Ryan and Kinder (1996a) suggest an event model which does emphasise anonymity. An area (a red light district) is selected and then either rejected or seen as 'acceptable' and then subsequently a specific location (brothel, or in the case of street prostitution, road) is then selected as protecting the sought for anonymity. Thus, while motivation to act may exist, it can be frustrated by the nature of the place. Obviously being away from home reduces the likelihood of the constraining variables being operative. Figure 2.2 is reproduced from the original text, but the definition of 'recourse to prostitutes' as 'unworthy' is now perceived as being problematical given the different layers of meaning

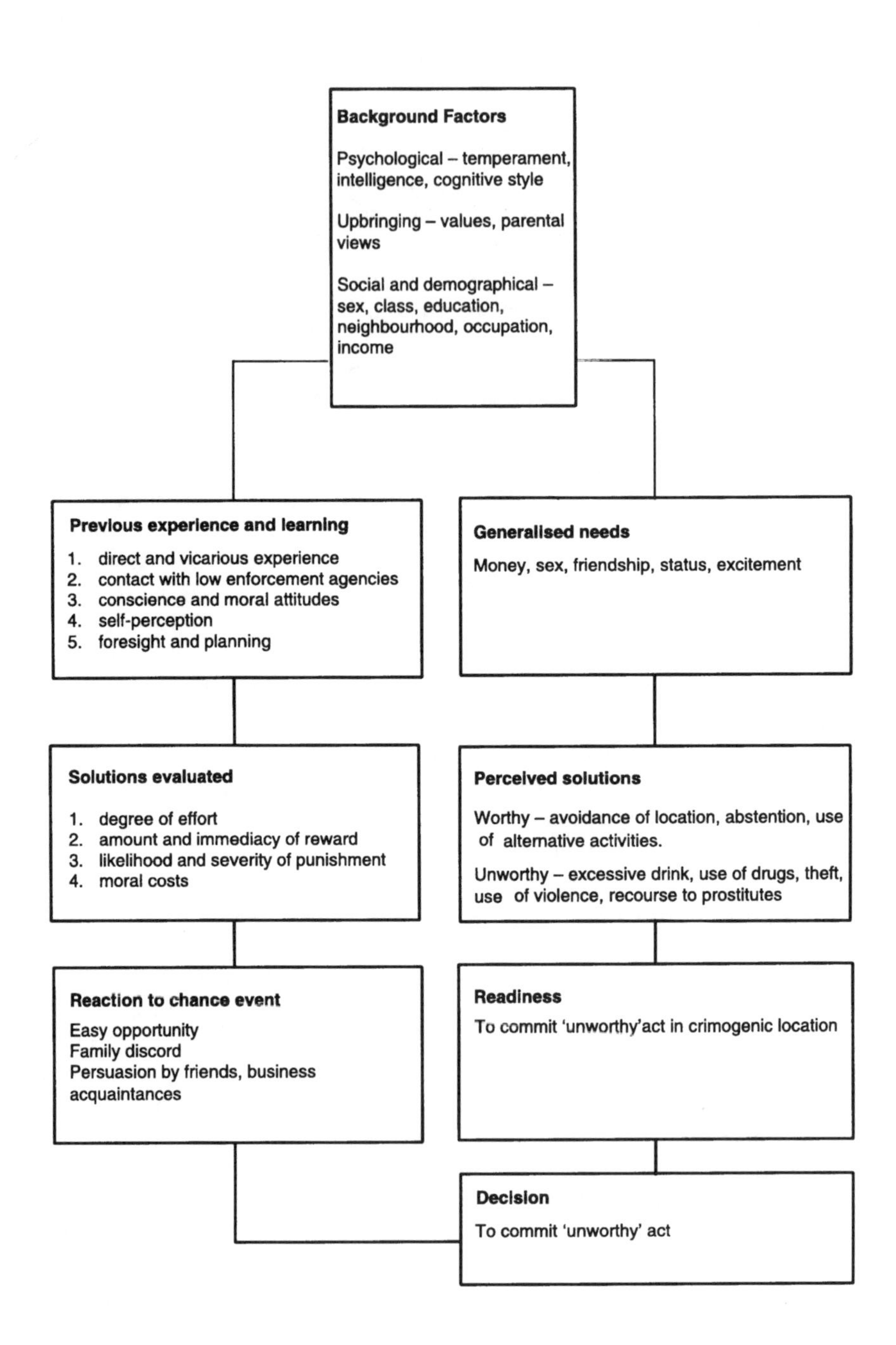

Figure 2.1 Initial involvement model

Source: Adapted from Cornish and Clarke (1986).

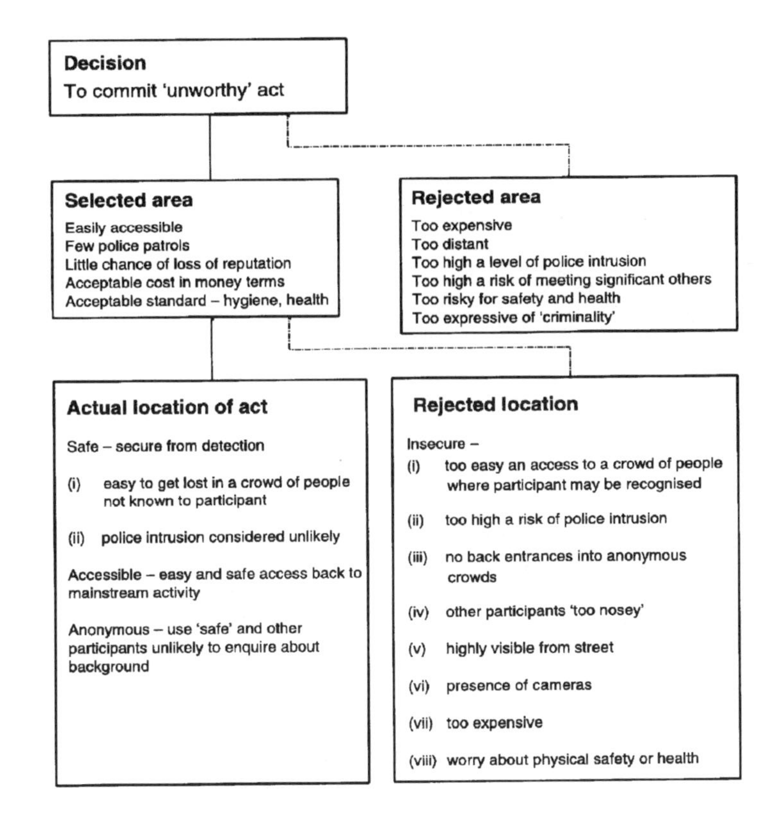

Figure 2.2 Event model
Source: Ryan and Kinder (1996a).

associated with commercial sex. Tonry and Morris (1985) link commercial sex with excessive drink, drugs, theft, etc., but this link is far from automatic, especially given the conditions of touristic destinations. Utilisation of the term 'unworthy' is here simply a reproduction of the original paper, but is not a term the present authors would automatically subscribe to.

The pragmatic or functional test of Turner's (1974) comparative symbology thus includes the spatial layout of the place (for example, see Ashworth *et al.*, 1988). The relationship between holiday motivation and that of visiting a sex worker is also symbiotic. As noted, both are leisure pursuits. The tourist trip provides opportunity. The tourist industry sustains the sex industry in part by enabling demand to be operationalised in locations away from home. As argued in Chapter 1, the tourism industry also sustains a pattern of global economics whereby the dominance of the one party ensures the economic dependency of the second. But a further pragmatic outcome is whether the interaction between the two parties has satisfied the needs of the sex worker and tourist, and this can be interpreted in terms of the result of the interaction upon the self-image of each. Such an analysis is consistent with the adoption of Turner's argument, previously noted that, tourist–sex worker interactions occur at a personal-psychological level. It thus appears that any analysis of sex tourism is operating at two levels of analysis. There is the micro-analysis of individuals who justify their actions, and the macro-level of analysis where the aggregation of individual acts is sustained by socio-political-economic matrices of interaction whereby some have monetary power to wield and others have no other way of accessing money other than to engage in prostitution. At the personal level there may be a choice to engage in sex work, limited choice or forced labour. The conditions under which the sex worker engages in sex work are important in shaping the personal-psychological level of interaction, and the consequent damage that might be done to people.

Hollinshead (1999) also identifies elements of tourism, which, for our purposes, can also be utilised to help explain the relationship between tourism and prostitution. He argues that tourists are drawn towards 'selectively celebrated sites', they delight in the 'exotic', and that they are self-indulgent. He also advances the notion that the 'tourist gaze' (Urry, 1990) is important because it helps render a more romantic and illusory world with the consequence that it transforms places and people, and creatively empowers 'the vision of people of and about the world' (Hollinshead, 1999: 5). From what has been described, these processes can be seen to be operating within sex tourism just as in other forms of tourism, as Hollinshead would no doubt contend would happen. Locations like Amsterdam, Patpong, or other red light districts attract clients and the tourist as voyeur, but in doing so the 'Gaze' itself acquires a political power based on a questioning as to why, how and should such places exist in the way in which they do. As Hoy notes, in the Foucaldian view of the world:

> Power in a game of chess is [not exercised] by one piece over another at the moment of capture. On Foucault's model, the capture is indeed a 'micro-power', but it is also the effect of the overall arrangement of the pieces at the time *as well as* of the strategy leading up to and including the capture.
>
> (1986: 135)

Thus tourism becomes not simply a provider and creator of demand for prostitution, but in the very processes of creating places attracts an attention whereby tourism becomes a catalyst for subsequent change. If Butler's (1980) destination life-cycle can apply to resorts in general, so too it might apply to specific red light districts. These issues are now examined in the next few chapters.

3 Paradigms of sex tourism

Thus far it has been assumed that the terms 'sex tourism' and 'prostitute' or 'sex worker' are self-evident, but it has also been shown that many different situations and interpretations exist. Hence, is there a single definition of these terms that encompasses all of these situations? It has also been argued that tourism presents an opportunity for people, male and female, to exploit their marginal status and their economic power to cross the line between the licit and illicit boundaries between the socially sanctioned and the 'socially suspect'. This blurring of the boundary needs to be examined more closely. The blurring or crossing of any boundary is not a neutral act. Nor are boundaries without importance. Simmel (1971: 353) notes: 'The boundary, above and below, is our means for finding direction in the infinite space of our worlds.' Thus, the redrawing of boundaries means a redrawing of the frameworks of personal and societal knowledge and values. Boundaries are points of disjuncture and flux, and thus perversely, the existence of a boundary may actually act to ease the transmission of discourse just as, according to Porter (1997) they act as funnels for flows of trade, HIV and sex trafficking. The crossing of such boundaries has the potential for the socially condemned, the apparent deviant, to enter the discourse of that which is normally socially sanctioned. The wider society must decide the basis upon which any penetration of the boundary with the 'illegitimate' is to be judged. The presence of, and interaction with, the illicit margin must be either denied or recognised. Various options exist. Either the illegitimate continues to be condemned, and those who interact with it are also subject to condemnation. Or, as is often the case with prostitution, the relationship between licit and illicit is left ambiguous, a grey area of hypocrisy, and as such subject to injustice. The ambiguity arises because of a wish to avoid punishment of the transgressor in the form of the client – a position that can only be possessed by one with power within the social framework, and whose transgression is thereby subject to varying degrees of toleration. Third, societal change may force a growing recognition of the injustices to which subordinate groups are subjected, and thus legal, political and social changes occur within which the mainstream society recognises those injustices and the conditions which created them. What causes this to occur? For any system both or either internal and external factors may force change.

In the case of sex tourism, as already noted, changing social norms within society as to sexuality may change views as to the roles that sex workers perform. It has been argued that within Western societies a greater recognition of the role gender plays, the separation of concepts of gender and sexuality, the growing tolerance of lesbian and homosexual communities and individuals, the changing roles of women and the expression of female aspirations through feminism have all had, as a by-product, a questioning of the role of sex workers. It is not a uniform discourse as has already been demonstrated, but it increasingly brings the role of sex workers, male and female, into a wider arena. The role of HIV/AIDS has also proven to be a catalyst in thinking. Evidence of this is provided through the personal experiences of those working with and on behalf of sex workers. Cheryl Overs, the NWSP (NetWork of Sex Projects) Co-ordinator and an organiser of the representation of sex workers at the 1995 Conference on Women in Beijing says:

> Everything has changed. The women's movement is no longer the home of the sex workers' rights movement. The happiest time for me was the late eighties and early nineties when HIV/AIDS brought new links with gay men, and Madonna rose and rose. Our small band of sex worker rights activists suddenly had a new political family. Once a pro-sex feminist theory was articulated we even had new supporters from the women's movement who were listening to sex workers for the first time. Young feminists matured with the notion of sex workers rights as a fixed entry on the women's rights agenda. I enjoyed that because I felt exonerated. It had been hurtful, as well as frustrating, that for many years feminist puritans have said that our demands for recognition of sex work as valid work were a product of false consciousness which blurred our perceptions of our damaging experiences as victims of the sex industry.
>
> (Kempadoo and Doezema, 1998: 205)

The importance of HIV/AIDS cannot be over-estimated in affecting the attitudes and policies of non-Western governments. Notably, the Thai government has had to recognise the potential spread of AIDS not only through its sex industry, but in the transmission of AIDS to the rural communities of Thailand. In July 1994 the Thai Ministry of Health estimated that the number of HIV positive people in Thailand numbered 650,000 (Seabrook, 1996: 121). While this represents about 1 per cent of the total population, the figures for the rural areas of the north from which many of the sex workers come indicate that 7 per cent of the population of those areas are HIV positive. Safe sex programmes are being initiated in the tourist bars (Hanson, 1998), but implementation appears to be uneven, and the problems go beyond the *soi* of Patpong to those areas which serve the need of local men. Safe sex programmes are also unlikely to have much impact on the traffickers of women. As will be described in Chapter 6, there does exist large-scale trafficking of women, much of it brutally exploitative, systematic and illustrative of a wider social ill within the body politic. That it occurs at all implies that corruption exists in the highest levels of government. That corruption is a problem with economic and social consequences has all too often been demonstrated. It is not a problem isolated to non-democratic governments. The evidence of experiences in Italy has shown that even Prime Ministers have been allegedly involved in corrupt practices. Organised crime in countries like Russia are known to be involved in prostitution and drugs, and the commodification of women under such circumstances is no longer an academic conceptualisation of spreading consumerism, but a harsh, dangerous reality for many women. Even within what are generally deemed to be more enlightened countries like New Zealand where women can work in licensed massage parlours, and where working as a prostitute is not illegal, danger still exists. In Chapter 2 reference was made to sex workers having, like any other worker, good days and bad. Unfortunately, as is known from past murders, the bad day might mean the loss of life. This real danger is even more to the fore when women are in positions of being abused.

Thus far, no definition of sex tourism has been undertaken apart from the notion that it involves interaction between liminal people. The point of any definition is to more clearly identify what is under discussion, but the very act of definition both focuses on and perhaps limits the nature of the problem. For this chapter definition will be attempted, but by means of first identifying the parameters under which sex tourism operates. From the first two chapters it appears that sex tourism operates under the following dichotomies from the perspectives of both parties:

1 It is 'voluntary' or exploitative.
2 It is commercial or non-commercial.
3 It is enhancing or degrading for self-identity.

While Ryan (2000) uses these three sets of continua to analyse paradigms of sex tourism, in themselves they are arguably incomplete. What is evident in the literature is that a macro- and micro-perspective exists. As already alluded to when describing Turner's conceptualisation of liminoid people, such an analysis as that suggested by Ryan (2000) tends to work best at the individual psychological level. As many authors testify (for example, O'Connell Davidson, 1995, 1998; Bishop and Robinson, 1998; Seabrook, 1996) when discussing sex tourism in locations like Thailand, there exists too a macro-level of socio-political structures which create and perpetuate the infra-structures of sex tourism on a global level. For O'Connell Davidson 'sex tourism must be recognised as first and foremost a form of economic exploitation' (1995: 61), but one unlike other forms of exploitation in that no mutual set of dependencies exist. She argues that sex tourists, as a collective group, are not locked into a dialectical relationship with Thai sex workers, as prostitutes are unable to win concessions from their clients who are free to move on elsewhere. The maintenance of the flow of sex tourists is, she argues, one that is sustained by 'governments, international travel companies, hoteliers, local business people and so on, who have an economic interest in maintaining the flow of sex tourists' (ibid.: 62). Thus, in her view, programmes that direct action against sex tourists have but a marginal chance of improving the lot of the women involved. To what extent such dialectical relationships exist is debatable, and thus it is important to examine each of the three paradigms listed above, but to do so at a micro- and macro-perspective.

To undertake such a task is, however, not one conducted in a positivistic frame of enquiry. There are multiple truths, and the three paradigms are not independent of each other, but interactive. Equally, the distinction between macro- and micro-levels of analysis becomes increasingly difficult to sustain when looked at in detail. It also raises the question of to what extent is it possible to develop an explanatory model of sex tourism? At the micro-level the interactions between sex worker and client are individualistic, even while commonalities may be discerned. At the macro-level the different legal regimes in various countries, the miscellaneous social and ethical *mores* – all contribute to a series of case studies within national settings as well as illustrating generalities. Thus, of necessity, sex tourism is pluralistic, multivocal and multi-dimensional in its forms.

Voluntary and commercial transactions

The question of how voluntary the work is from the viewpoint of the sex worker is a moot point. It may be argued that no sex worker enters the industry on a voluntary basis as it is a choice made generally out of economic necessity. This is not a new observation. Bell (1994: 103) cites Parent-Duchatelet's study of Parisian prostitutes in 1831 and the conclusion that: 'Of all the causes of prostitution, particularly in Paris and probably in all large cities, none is more active than lack of work and the misery which is the inevitable result of insufficient wages.' There is sufficient evidence that this continues to be the case from what has been published on prostitution to mention McLeod (1982), Boyle (1994), Jordan (1991) Kempadoo and Doezema (1998) among many such writers. It is seemingly a motive related to many forms of sex work. Ryan and Martin (2001) cite such a motive among many who enter the sex industry as strippers or exotic dancers. For example, one of their Australian respondents stated: 'If I was an unskilled man, no qualifications, left school early, I could still get $30 an hour by going on the building sites. What does a woman do in the same position – we can't do the same.'

Another, who had been an exotic dancer for ten years, described how she had left home, had initially had a job working in a retail store, had left that job because she had had a disagreement with a supervisor and thus:

> There I was, no job, no money, so I started stripping. The only reason why you do it is the money. I know I can't keep doing this so at the end of this season I intend going back to school – I'd like to do psychology – you learn a lot in this job.

Yet, while there is money to be made within the sex industry, few women will become rich, or even specifically well off from the industry. Prostitution may provide 'the bread', but the extra 'jam' is more noticeable by its absence rather than presence. Adult film actress, Nina Hartley, makes the point that 'while it's not a bad living, you ain't gonna get rich doing it', and points out that, as a film actress, you have to understand what it means to 'have a permanent record of you with a man's penis in your mouth' (Chapkis, 1997: 35). The monetary expenses of working in the sex industry are high. Ryan and Martin (2001) describe the expenses involved in exotic dancing – the high cost of shoes which wear out quickly and the need to continually prepare costumes. It is of interest to note the different interpretations of the same act. What to the sex worker is a necessary investment is, for the moralist, a sign of decadence. To again cite Parent-Duchatelet, among the reasons cited for prostitution are 'vanity and the desire for the glitter of luxurious clothes is with idleness one of the most active causes of prostitution, particularly in Paris' (cited by Bell, 1994: 99).

While sex workers cite money as the reason for their work, like all workers, it is not so much money *per se*, but what money permits that is the reason for their work. Many motives and wants exist. For some, money is the way in which a decent lifestyle is obtained. Sex work is the means by which single women provide food, a home and clothes for their families. For some older women, it is a means to bring in extra comforts. For others, as described by Nina Lopez Jones, it may be a passport to safety. Thus:

> Having the money … is also a way of avoiding a lot of other rape and violence in your life generally, whether from your family or not. It means you can afford to take a cab rather than waiting out on the street for a bus and running the risk of being attacked there. It means that you don't have to work necessarily in a job where either sex with your boss, or some kind of sexual work with your boss, is a requirement. It means you don't have to sleep with your landlord so that he doesn't evict you.
>
> (cited by Silver, 1993: 106)

Plumridge's work in New Zealand seconds this as a motive. Reporting this research Ansley cites one respondent as saying:

> I thought you got married, you'd be bright-eyed and bushy-tailed and rosy down the track. Instead there were 15 years of hell, beaten black and blue, raped … I went through hell, broken windows, baseball batted. To him I was nothing but a sponging whore.
>
> (2000: 22)

Yet Plumridge's work confirms that of other researchers that, in New Zealand, as elsewhere there exists a hierarchy of prostitution, and while parlour workers may prosper, those who work on the street continue to risk being beaten. The escape from violent marriages, for a minority, has so conditioned them to the prevalence of violence that the ever-present risk of danger for those working on the street continues to be less than that previously experienced.

But other motives also exist for continuing in the sex industry, either as prostitutes, strippers or workers in the pornographic industry. Ansley (2000) reports that the motive to work as a sex worker is not always born of a wish to escape financial marginality – it is a means by which women can prosper. It is a way of getting ahead, of getting a house. Nonetheless, such advancement does depend, as already noted, upon the management of the brothels where the women work. Bonds may be required, fines levied, and when business is slow, charges may be introduced for things like laundry bills – all of which reduces the money earned. Yet several women do find it an interesting and reasonably lucrative lifestyle. In addition to money, the work brings several advantages of being an independent worker able to travel and work in many places of the world quite easily. Ryan *et al.* cite one of their respondents, an Australian worker in a Wellington, New Zealand, massage parlour, as saying:

> At present I'm working here – I'll do it a for a few weeks, then I'll go down to see South Island and work in Christchurch – and that way I'll see South Island – I'll stay as long as I want. I've worked my way round lots of countries – it's a really good way to see places – you can stay as long as you want, people are nice – its better than most jobs.
>
> (1998: 323)

Another sex worker said: 'It's a job – but it's not a bad job! I can work in a parlour or private. I've worked in Oz – had a good time there and spent longer and saw more than if I'd been in any other kind of work.'

In Ryan and Martin's (2001) paper on exotic dancing and the tourism industry one respondent, who had been dancing for eight years said:

> I had been working in Woolworth's for a year and thought, I want to do more than this. I saw an advert for a club wanting bar work, so I phoned up, and it was quite open that it was a strip club, so I thought why not. Anyway I went for an interview and I walk in, it was the xxxx in Melbourne, and they have low tables for table dancing, and there was this women lap dancing and sticking her butt right in this guy's face and another dancing on the table and showing her pussy to this guy, no hands hiding it you know, just full on. I'd never seen this before. I was really naive, just 18, and I thought could I handle this? But I didn't want Woolworth's and so I got the job. At first I hated the girls, I thought all they wanted was just money. I didn't talk to them, but eventually I did and that's where I met xxx and I got to know just how nice they were. So eventually did a bit, you know never done any before, didn't know how to dance or handle the bar, and I got $2000 one week and thought 'wow'. Anyway, I decided to travel, hadn't been overseas before and went to London – that was great!

This statement confirms the motives for the need for money, but it also raises the issue of how women come to terms with the raw physical presence of sex work. O'Connell Davidson (1998) describes the ritualistic use of prostitutes by groups of men seeking to initiate one of their group into sexual experience. What is it like, she asks:

> to be taken to a total stranger by your friends or relatives and expected to immediately undress and engage in penetrative sex with this unknown person? What would it feel like to be expected to publicly and anonymously do that which you have been taught to believe is both private and intimate?

But it can be argued that the reality of the purchased sexual act for women is one of continually having to cope with this physical presence and question. What is it like to continually have to undress for a client whom you may not find attractive, for someone who by the power of the wallet alone has a perceived right to penetrate you, to fondle your private parts, to possibly enable them to live a fantasy that they would not otherwise talk about, much less do? And for the sex tourist, who in addition to the normal demands of a local client may bring a baggage of racist stereotypes of wild black women or demure oriental women, how does the woman cope with these demands not just occasionally, but perhaps every day of her working life? For O'Connell Davidson (1998: 141) the answers partly lie in a process of commoditisation where the one creates of the other 'a function, an object' and where prostitution 'is valued because it strips women of the autonomy and separateness which clients find so very threatening' (ibid.: 154).

Yet, as has been already discussed, the very process of commodification permits the development of control, and of professionalism on the part of many sex workers. This process of commodification has numerous sides to it. First, as will be discussed more in the next chapter, it contains the danger of the sex worker perceiving him or herself as a separate commodity in order to survive in the job. In an interview with the first author, Jo said:

> Why do women become engaged in sex work? It is the dressing up, looking good, being complimented, and it is the discovery of just how easy it is. Three minutes, and all that money. But after a time the hanging around, the junk food, the continued routine just gets to you, and then it's the time to get out. There are a lot of women who are in it too long, they get hooked on the money, and they get hooked on drugs. The drugs are a recreational thing, and they help you get through the job. And there is a culture of drug taking – everyone is doing it.

From Jo's perspective, as a sex worker and in her role as an outreach worker, it is when you separate this commoditised role from yourself, and the role is the 'other', then it becomes necessary to leave the industry as this psychological distinction is but a short-term mode of coping and in the longer term is dangerous to the psyche. Michelle, another outreach worker and former prostitute stated:

> Those who take on the role of the 'priestess' are so very few – they are a minority. Equally those being exploited by men in relationships are also a minority – they are extremes of a continuum, most fall in between. Each of us works out our own way – yet – I think I am a fully mature woman, able to cope with most things, but sometimes things come up from my past that are difficult to cope with – so what is it like for others?

Thus, if sex work is a commodity, it is also a job, and it is job within which women are able to, and do, take pride. The recognition of sex work as a job also means that the prostitute is no longer working at the margins of society, but can become a recognised tax payer and contributor to society through the normal ways in which any employee or employer does. Thus, for example, the New Zealand Prostitutes Collective has given advice to sex workers as to what job-related expenses can be claimed against taxation (Siren, 1998). However, what is applicable in New Zealand is not applicable in every country. Thus Phoenix (1995: 66–67) argues that if the nature of prostitution is to be defined it must take into account:

- Place of work: where does the work take place?
- Mode of client contact: how do clients and workers get into contact with each other?
- Employment status: is the worker full-time, part-time, regular or casual?
- Peripheral activities: are there any other activities commercialised in the prostitution exchanges?
- Exchange practices: how are monies collected and distributed? What types of contracts exist?
- Formal relationships with others – what relationships exist with others apart from the client?
- Risks and protection: what risks to the worker are involved?

Thus, there is considerable difference between the girls and woman taken by force to work in backstreet brothels where she may be little more than a prisoner being raped by men who make payment only to her captors, to one who works independently, pays tax and who has considerable control over her clientele. Phoenix argues that significant differences exist between those who work on the streets, and those in massage parlours, escort agencies or from their homes. There exists significant evidence that there are considerable differences between those who work in the same country as to working habits and patterns. For example, Plant (1997) argues that distinct differences exist between drug taking by sex workers in Glasgow and Edinburgh, noting very high levels of intravenous drug users in the former case whereas, in the capital, drug usage appeared to be of alcohol and cannabis. Pimping was much higher in Glasgow than in Edinburgh. The existence of pimping certainly seems to reduce any degree of voluntary work on the part of the sex worker. Silbert and Pines (1982) interviewed 200 sex workers in San Francisco. Seventy per cent reported that various forms of abuse affected their decision to become a prostitute and once on the streets, 70 per cent had been victimised by clients and pimps. Faugier (1994) argues that women in relationships with pimps are for the most part damaged both physically and psychologically. Milner and Milner (1972) quote one pimp as describing his relationship with a female prostitute as one where her personality is changed 'You create a different environment. It's a brainwashing process; the whole thing is creativity. When you turn a chick out, you take away every set of values and morality she had previously and create a different environment' (Milner and Milner, cited by Scambler and Scambler, 1997: 123).

The presentation of the pimp as an evil manipulator should not, argue Høigård and Finstad (1992), lead to 'easy' representations of the pimp as 'other' on the basis that it demarcates prostitution and its ills as a problem separate from wider society. Among many observations they make one might note their comment that: 'Many prostitutes have resources and personal qualities that make the average academic seem colourless and weak. Being in the position of a victim is a structural and relational attribute, separate from individual characteristics' (ibid.: 183), and 'When the image of the outsider is not blocking our perception, we have to recognise that a number of young women today experience prostitution as the least of numerous evils. This says a good deal about their other alternatives' (ibid.: 207).

As seen from these examples and others previously discussed, there exists a considerable literature which has analysed the position of the prostitute as being the result of social structures that are male-dominated and possessing inequalities of economic wealth and power. It is argued that the sex worker is subordinate at both a personal and structural level because of male demand, and because males generally occupy the higher income levels. However, at least two different viewpoints exist. It should be noted that to argue different viewpoints exist is not to state that they are contrary perspectives, as the nature of prostitution and sex tourism is such that simultaneous differences can co-exist.

First, there is the issue that at an individual level prostitution does permit an expression of economic and pyschological independence. As already noted, it offers an occupation where women may decide their working hours to fit around their other roles as mother and care-giver to dependent others. While it may be argued that it is a structural problem that few well-paid jobs exist for women without high educational qualifications, it is true that sex work does provide such an occupation. Some feminists have argued that all that the prostitute does is to make explicit a truth about the female condition – that marriage itself is a contract whereby the woman exchanges sexual services for financial security. For Miles (1992) men get married only because they have to as a social obligation, and society rewards them to ensure their compliance by endowing them with power within the marriage. Yet, she continues to argue, the state of marriage is such that it often ends up emasculating both parties. It is also a misrepresentation to argue that all sex workers have low levels of educational attainment. Equally, it is a falsehood to argue that all those students who do enter sex work to pay for tertiary education actually give up sex work upon graduation. Equally, it is false to maintain that all such students feel constrained in subsequent career choice because of a fear that their career will be prejudiced by their past sex work. Nonetheless, to argue that most sex workers are well educated would be an untenable position.

There are sex workers who admit to enjoying their work and their relationships with clients. Jordan (1991) interviews one such worker who claims to love all her clients. In interviews in Siren seven years later the same woman was still able to write:

> To all those clients that I have loved over the past 18 years I say thank you for those stolen magic moments … You have helped me understand the invisible thread between my being and yours, to know that sex is far more than just a physical act but a dance of our souls together, unique and precious.
>
> (Siren, 1998: 6)

However, the same issue contains another interview that began: 'Sex work sucks! Clients suck! (or hassle me to try and have a go at sucking), I wish I could tell the whole business to go suck! But, and it is a big BUT, I want the money! Who doesn't?' (ibid.: 7).

Another former sex worker wrote:

> I had a wonderful joyous time my first two years in this business. I gave each customer my full heart for that time I was with them. I loved each and every one of them. I had fun dressing up and felt light-hearted, happy and free and couldn't understand why the girls who had been in this business longer were so negative and depressed all the time.
>
> Then after two years, it started to hit me too. I found myself being an actress more and more and not coming from the heart. Money became foremost and important. The fun was gone.
>
> (Sisters of the Heart, 1997: 104–105)

This need for 'time out' has been stated several times to the author by different women. To again quote Michelle, 'It is important to take time out, you can't keep doing it – you need time out to heal yourself.'

While being prepared to accept both versions ('I work because I like it' and 'I work for the money') as possessing truths for their respective respondents, the client would much rather accept the former than the latter. The sex tourist wishes to engage in a belief that the act is a voluntary one, or if not voluntary, one that is permitted by local value systems. Thus those who comment upon the sex industry of Thailand note how often Western men (at least) argue that their relationship with Thai women is not a simple contractual one based upon an exchange of money for sexual services. Günther (1998) cites one sex tourist to Thailand who viewed the relationship as a romantic one, even though the woman concerned worked as a prostitute. 'R' is not a sex tourist because he lacked intent, he did not restrict his vacation to having sex with local women, he lacked promiscuity because he did not have sex with several women, he did not allow his sexual impulses to control his behaviour without constraint, and, according to Günther, for his respondent, the most telling argument is that he did not pay. Yet payments were paid, including bar fees. As is discussed by Cohen (1993), Seabrook (1996), O'Connell Davidson (1995, 1998), Bishop and Robinson (1998) among other researchers, for many sex tourists there is a wish to deny their role as sex tourists. Equally, if that role is recognised, then it is matched by a wish to impute to the women those motivations that are simply mercenary, and furthermore, manipulative.

The attitude of the sex tourist has been variously identified as being exploitative and exploited, depending in part upon the writer. The contributions by sex tourists to the debate have been primarily through their discussions with academics (Høigård and Finstad, 1992; O'Connell Davidson, 1995; Seabrook, 1996; Ryan and Kinder, 1996b; Bishop and Robinson, 1998) or through publications generally aimed at other sex tourists, either in conventional paper form or on the World Wide Web. A possible intermediate level is expressed in fiction written by those with experience of the bar scene of places like Bangkok. An example of this latter is the novels of Christopher Moore (1991, 1993, 1995) with its insider references to 'HQ', 'termites' and other slang of the area. Many of Moore's characters reiterate the mantra of the Dexter-Horn video productions that Western women are demanding, ungiving, and hence the world of Patpong's bars is really the natural world where men can be men, and women are women who look after men. In *A Haunting Smile*, one of Moore's characters states that 'fucking a white woman is a step away from homosexuality' (1993: 107). This inability to deal with women as other than vaginas awaiting the thrust of the penis is a common feature of the interviews reported by O'Connell Davidson (1995) and Seabrook (1996). This comment also draws attention to the racial stereotyping that is prevalent in a number of the comments of clients that are reported by researchers. Cabezas (1999: 111) reports that Dominican women reported 'that foreigners constructed them, both sexually and racially, in opposition to European women. Their amigos portrayed European women as cold, indifferent to sex, and like men at home.' Such portrayals reflect as already discussed, the mix of racial exoticism, the inability to cope with 'liberated women', and constructs of 'womanly women' and 'male men' by reference to norms derived from some perception of nineteenth-century values as described in Chapter 1.

However, one problem with such reported conversations is that samples are usually small in number, and while it is not doubted that such views are expressed, it also needs to be recognised that the ontological expression of the research is that of critical reality. Guba (1990) defines this as research that is ideologically oriented inquiry. Its epistemology is subjectivist in the sense that values mediate the inquiry, and the methodological approach can be characterised as dialogic and transformative. The views of the sex tourist just enunciated are consistent with the view that 'sex tourists express a kind of misogynistic rage against women who have the power to demand anything at all, whether it is the right to have a say over who they have sex with and when, or the right to maintenance payments for their children' (O'Connell Davidson, 1995: 53).

In an analysis of Christopher Moore's series of books set in the bars of Patpong, Bishop and Robinson (1998: 171) discuss the 'Aging, alienation, and intimations of mortality' that characterise the 'hardcore' client. They note that the sex tourists, so caught up in their own world of the fantastical, argue that it is they, the clients, who are the victims, not the girls that they use. Such, argue Bishop and Robinson, is the extent of self-delusion and misplaced values that the sex tourist espouses.

As a result of his own research this author does not doubt the ability for self-deception that exists within human nature, and equally does not doubt the role of the researcher as a catalyst for sex tourists sometimes having to face these self-deceits. Yet, the rage, the inability to face truths, the disillusionment that is experienced when men realise that for Thai women sex work is but a means of economic support, little more, and thus the sought for affection may have been little more than a skilful play – the resulting mix of emotions may nonetheless be real. Sex tourism for some males becomes like a drug. They repeat their visits, they spend their money, they repeat their cycles of disillusionment. From one perspective the irony is that in an attempt to continually enforce their concepts of masculinity though the roundabout of visits to bars and massage parlours, the men observed in this cycle continually re-state their weakness. For a writer like Schlessinger (1997), men are characterised by a 'stupid strength' and an inability to discuss things beyond a superficial level. Men, she writes, are uncomfortable with feeling weak, useless or rejected, and react by intimidation and use of force. Maleness is measured by sexual conquest. From this perspective the male sex tourist is little more than a man unable to come to terms with himself, and in that inability is running true to his being.

But if this is true of males, what of the female sex tourist? And, is all sex tourism like that of Patpong? The answer to the latter question is 'no' if one looks only at the volume of the sex business and the nature of its presentation. Also, as Hanson (1998) reminds us, the sex tourism engaged in by the white *farangs*, high profile that it has in Western literature, is only but a mere fraction of the totality of prostitution, and, indeed, sex tourism in Thailand. For example, until the Malaysian government stopped the exchangeability of its currency in 1998, one of the main locations for sex tourism in Thailand was that of the Malay–Thailand border. The recognition of female sex tourists also raises a number of issues relating to the voluntary–involuntary nature of sex tourism. Albuquerque (1998) identifies four types of female sex tourist, namely

- the first timers – the 'neophytes';
- the situational sex tourists, who do not travel with the specific intention of buying sex, but avail themselves of the opportunity when it arises;
- the 'veterans' who travel explicitly for anonymous sex and usually find multiple partners; and
- the 'returnee' who visits to be specifically with one man whom she has met on a previous visit.

For her part, Phillips (1999: 190–191) identifies three types of female sex tourist, 'The Situationer' who emphasises romance, the 'Repeat Situationer' who too denies the remunerative nature of the relationship, and 'The One Nighters' who come for fun and to 'fuck a black man'. It is immediately obvious that such categorisations can apply to their male counterparts. If this is the case, do the same descriptions of male sex tourists as self-deluding, exploitative and weak apply to their female counterparts? Certainly, Albuquerque makes an explicit parallel in the case of the 'veterans', whom he compares with O'Connell Davidson's 'macho men' – in short, the transactions are explicit and focused, they 'find them, feed them, fuck them and forget them' (Albuquerque, 1998: 90). Likewise, the 'situational sex tourist' is akin to Günther's (1998) 'R'. They do not intend to be sex tourists and she sees herself as a patron in terms of providing financial support, she tends to be loyal to the 'boy' for the duration for the holiday and sees the relationship as genuine and reciprocal. Companionship and a sense of romance are as important as the sex. Thus too with the 'returnee'. She has romantic sentiments, is loyal to the male, and brings back gifts. The Barbadian 'beach boys' who form part of Albuquerque's sample have the same behaviours as the Thai female sex workers described above. They work out schedules around visiting girl friends, they are not loyal to one foreigner. Indeed, it is observed that their local girl friends have to 'share' their boy friends with the visiting sex tourist. But Albuquerque observes that many of the women are just as promiscuous as their male friends and will abandon their beach boy if another takes their fancy.

However, one partial difference is played out, and that is with the possible exception of the 'veteran' it is the male who tends to initiate the procedure. Yet even the 'veterans' have to play by the rules, which are that too much public display of a sexual nature by female tourist (other than on the dance floor) is seen as being in bad taste. The role of fantasy is also evident. In the case of male sex tourists to Thailand it is seemingly the search for the compliant, submissive female. For women travelling to the West Indies, it is a case of stereotypical sexy black males who are active 'studs'. However, Hosein's (1995) work with young Caribbean women finds a local female opinion that Caribbean males are poor lovers, unskilled and simply interested in a 'slam, bam, approach'. On the other hand, Phillips (pers. comm.) describes the specific strategies used by Caribbean beach boys. They tend to select women who are not tanned (thereby implying that they are recent arrivals). They seek women who may be a little overweight on the premise that they may not often attract men and so be more susceptible to their charms. Thus, like O'Connell Davidson's pictures of males who are balding, ageing and overweight, so too a similar picture might be painted of the female sex tourist. Just as Thai bar girls target their 'partners', so too do beach boys. The parallels and complexities of who is exploited and who exploits exist regardless of the gender of the sex tourist; and equally it may be said, any picture painted of these complexities says much of the researcher as well as the researched. Are, it may be asked, all sex tourists really so physically unappealing?

The Caribbean beach boys (Albuquerque, 1998, 1999), the Palestinian shopkeepers (Cohen, 1971; Bowman, 1988), the Gambian males (Yamba, 1988), Greek males (Wickens, 1997) and other groups of males around the world meet the needs of female tourists seeking sex and derive payments from it. But the nature of the market transaction is a veiled one. It is veiled because there is no payment of a previously agreed price. Through the money veil of payment for meals and drinks, through the giving of gifts, and in many cases through a sentimental attachment between both parties, however transient, similar needs are being met as discussed in the case of many of the male tourists to locations like Thailand. The processes of exploitation in the sense that one can afford the gift giving and the other requires the gift giving to sustain a desired lifestyle is still present, but exists within degrees acceptable to both parties.

If these sets of relationships are perceived as sex tourism, then does the money veil also come into play when neither party to the transaction is a veteran at the sex game? There are significant levels of evidence to suggest that many tourists, both male and female, are not adverse to sexual adventure if it happens during a holiday. Clark and Clift (1996) found within their sample of British tourists in Malta that 7.7 per cent admitted to having 'a romantic relationship' while on holiday. Gillies and Slack (1996), in a sample of 541 holidaymakers, found that 5 per cent had had sexual relationships with other than their normal partner while on holiday. Ford and Eiser (1996) found that 24 per cent of their sample of 1,033 people under the age of 29 holidaying in Torquay had sexual intercourse with a new partner. Ryan and Robertson (1997) found that 12.5 per cent of their student sample of 400 had sexual intercourse with new partners while on holiday. Are all of these people sex tourists?

At one level, the answer has to be 'yes'. For many of them the possibility of sex while away from home had crossed their mind. Ryan and Robertson (1997) found that among the items taken on holiday about 10 per cent packed condoms along with the sun tan lotion. An intent existed. However, for younger people the patterns of behaviour exhibited on certain types of holidays that revolve around bars, clubs and partying was very similar to normal weekend activities undertaken in their home town. The activities, including sex, were the same, only the location differed. In this respect it is of interest to note that one Auckland massage parlour, in 1998, while advertising for new female staff, contained within its advertisement the message, 'think of it as a night out with the girls'.

At the micro-level of interpersonal relations it seems that while we concentrate on the icons of sex tourism that are evidenced in the bars of places like Patpong, we overlook the social structures that increasingly make such locations not a marginal place that is 'over there', but part of a multiplicity of sexual relationships that exist within wider society as a whole. Sex is increasingly more apparent, more discussed, more a part of advertising and of popular culture. Raw sex is part of a culture of self-expression, rebellion, acceptance and rejection. Skeggs (1994) quotes the lyrics of the pop song 'Two Minute Brother' sung by Bytches with Problems:

> Is this all you got?
> one minute and you go pop
> yous is a big disgrace …
>
> I hate guys who talk a lot of shit
> how they last long and got good dicks
> talking shit and telling me lies … the best lover?
> they all two minute brothers

The challenge to male performance is explicit. Female black rappers may represent one extreme of a continuum between assertive and passive female roles, but they are representative of many females today of all ages in making distinctions between overtly expressing sexuality and being available for sexual usage. The role of the body, the expression of its sexuality and its relationship to identity are discussed in the next chapter, but it is worth also noting that the expression of sexuality and sex tourism of a new sort emerged in the latter part of the twentieth century and is becoming more important in the early part of the twenty-first century. These new forms of sex tourism relate to female demand and the emergence of gay and lesbian cultures in Western society, and their importance relates not only to the economic implications of such demands, but also to their impact upon the nature of the discourse about sex, sexuality and its recreational uses.

Before discussing these issues, it is perhaps pertinent to summarise the discussion thus far. Three dimensions have been suggested as being important in arriving at definitions of sex tourism. Thus far the discussion has concentrated on issues relating to problems of how voluntary participation in sex tourism by the sex worker (male and female) is and the level of commerciality involved in the transaction. This can be represented by a diagram in which one line represents a continuum between voluntary participation on the part of the sex worker at one extreme, and at the other, a position of total exploitation. A second axis represents a continuum between the purely commercial transaction and at the opposite an interaction not marked by set prices and menus of prices. (For such a menu see Sisters of the Heart, 1997: 41–48.)

It then becomes possible to locate various forms of sexual interaction on these axes. A suggested perceptual map is shown in Figure 3.1. The purpose of Figure 3.1 is not to provide a definition of locations of various forms of sex

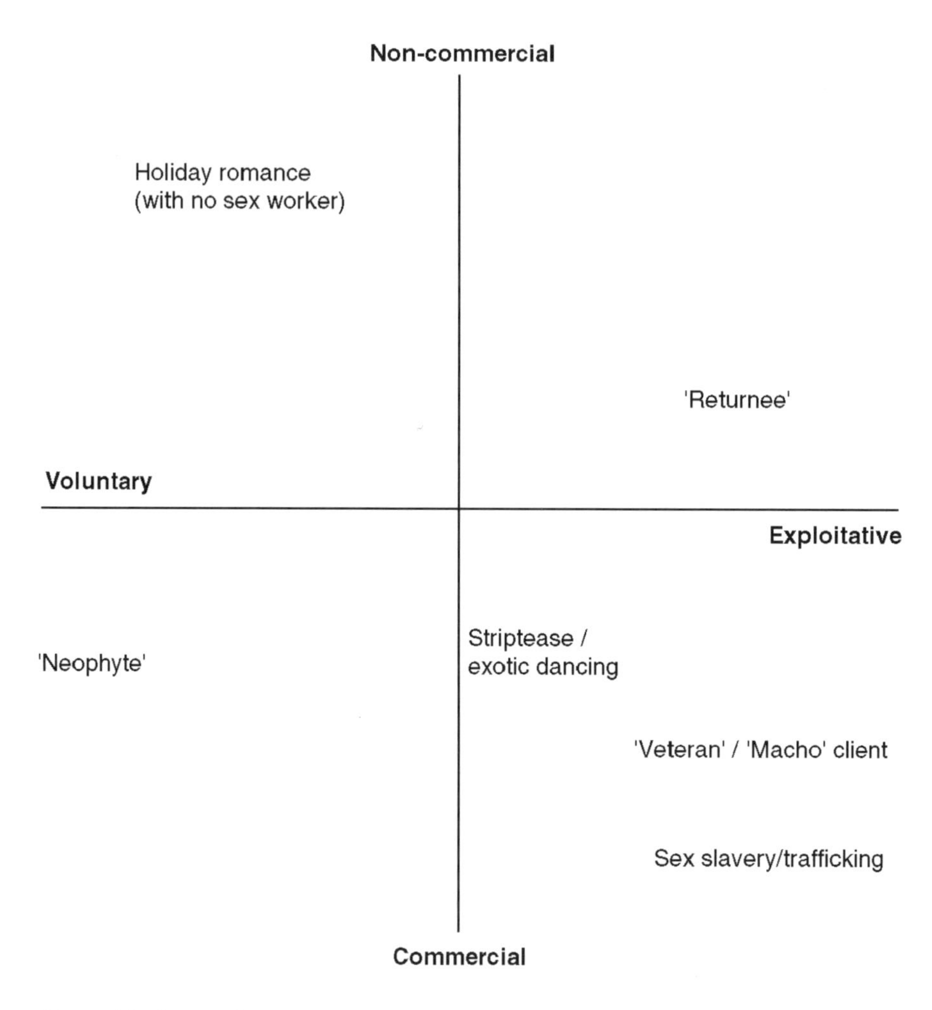

Figure 3.1 Sex tourism encounters

tourists–sex worker interaction, but to show how fluid the actual conceptualisations are. While the holiday romance between two people, neither of whom is a sex worker, appears a clear example of a voluntary, non-commercial interaction, what is the situation when one takes into account the small proportion of people who have more than one partner on their holidays? An intent for sexual adventure thus exists, and it can be argued to be exploitative – indeed, more exploitative than is the case involving the sex worker. It can be exploitative because:

1 no payment is made to the partner;
2 there may be instances where the partner is perhaps not entirely happy about engaging in full sexual intercourse. This permits a range of scenarios

> from rape to one where, while consenting at the time, one partner may subsequently suffer feelings of guilt, remorse, or of 'being used'.

Ryan and Robertson (1997) draw attention to the role of alcohol in sexual encounters among young people while on holiday, and thus the degree of collusion between the 'non-commercial parties' is thus debatable.

Under these circumstances the location of the 'holiday romance' type of encounter must move to the right in Figure 3.1. The position of the 'Neophyte' is placed in a voluntary setting as it is the sex worker who initiates the 'chat up' process and selects the client. This can also happen in certain types of sex work settings. In some of the clubs in Darlinghurst, Sydney's red light district, while the club may be a strip club, women will circulate among customers and ask if 'extras' are wanted. The 'extra' may be a lap dance, but the nature of the interaction takes place away from the public arena. The sex worker has a degree of choice as to which client is selected for an invitation, and the nature of the invitation being made. If she does not want full sex, then the offer being made will be that consistent with the nature of the club being a striptease club. The 'returnee' occupies an ambiguous location. The nature of payments is non-commercial in the sense that there is no menu of prices and payments are in the way of gifts and daily purchases, and it is exploitative in the sense that the professional sex worker may take advantage of the client, even while the client is occupying the economic position of being able to purchase affection.

The 'returnee's' position illustrates the micro- and macro-dimensions of sex tourism, and the ambiguous nature of the boundary between these two aspects. At a micro-level the actual interaction may be psychologically fulfilling to both parties, and genuine affection may be present. Both parties enjoy each other's company, and the parting may be associated with genuine feelings of sorrow and joy about future meetings. Yet, at the macro-level, the personal interaction perpetuates a system of economic dependency on sex tourism, it may do little to offer the sex worker alternative ways of earning a living, it illustrates the economic domination–subordination inherent in different levels of income – in short, it reinforces the very global systems of which commentators like Seabrook (1996) and Bishop and Robinson (1998) complain. However, within this relationship it becomes very possible for the sex tourist to reject the term 'sex tourist' and to argue that compared to the philandering holidaymaker with several non-commercial sex partners, theirs is a less exploitative position.

Possibly the least problematic aspect of sex tourism, from a definitional viewpoint, is that associated with 'veterans', 'macho' clients and sex slavery. For the former two categories of sex tourism the whole transaction is one of a commercial transaction purchased through the possession of economic power. However, as will be discussed, whether it can be fully regarded as an exploitative transaction can still be questioned. The problem of Figure 3.1 is that while it seeks to define paradigms of sex tourism, it does so by using ill-defined terms of 'commercial/non-commercial' and 'voluntary/exploitative'. Can one be exploited if one does not feel as if one is exploited? For radical feminists like

Dworkin, Barry and Marcovich the answer is that all sex workers by definition are exploited – every act of penetration is an abuse of power, whether the sex worker realises it or not. However, that is not a view shared by all, especially by many men and women who work in various roles in the sex industry.

'Sex slavery' seems to be an undoubted example of the worse kind of exploitative commercial relations. As such, it is discussed in much greater detail in Chapter 5 and arguably it lies outside of the conceptualisation shown in Figure 3.1 in that the very severity of degradation and abuse involved distinguishes it from the other situations being described.

Striptease is not often included in discussions of sex tourism, although Oppermann's (1998) book includes two chapters relating to striptease, but the discussion is more one of the exotic dance as a recreational activity. Ryan and Martin (2001) argue, based on an analysis on a six-month period in a club in the Northern Territory, that a significant part of the clientele are tourists, and this is even more true of locations like the Gold Coast in Eastern Australia or cities like Las Vegas. Equally, the dancers themselves are often travellers if not tourists, using their occupation to travel around their countries or overseas. In the case of Sydney's red light district, several of the Kings Cross exotic dancing venues prefer to hire trained dancers as they are able to provide a more interesting and sophisticated show to their audience which may keep them in the bars longer. Furthermore, such exotic dance venues are usually regarded by the dancers themselves as good, safe money which enables them to dance to an audience and keep fit. Nevertheless, as a form of sex work it raises many of the same issues of identity, sexuality and roles as prostitution, and the boundaries between striptease and other forms of sex work are often easily infiltrated. The strength of these boundaries often lies in the self-perception of the dancer and what it is that she or he wishes to do. Yet exotic dancing does differ from other sexual interactions in the sense that it places the body as an aesthetic experience in a non-contact way and as being on show in a public arena in a manner arguably not associated with other forms of dance or sex work. It is difficult to locate within Figure 3.1 and has been located centrally but with an inclination to the commercial. It is commercial in the sense that it is an occupation engaged in for money. That very fact reflects the lack of opportunities for some young people, yet often strippers have an interest in dance as a valid form of artistic endeavour, and indeed in exotic dancing as having, even if rooted in burlesque, a long theatrical tradition that is not without honour (Dragu and Harrison, 1989; Ryan and Martin, 2001).

Yet even in the case of exotic dancing, the boundaries are blurred. Thus in an interview with an exotic dancer, Dominque said:

> An exotic dancer is one who knows how to dance – a stripper can just stand there and go (shakes her shoulders) just showing a great pair of tits and they can make money that way, but an exotic dancer can dance – its a show. When we do the 'fantasy room' that's just stripping, it's so small that you can't really dance. But you see the whole lot there, you know, full view of

> pussy. How much is a prostitute? I had this guy in and its $70 for a prostitute, and we get $50 for a 5-minute show – it's strange!

But lap dancing does involve contact and one stripper noted that:

> What's the difference between stripping and prostitution? If I lap dance and a guy comes because I think what the hell, does it make any difference if I was doing a hand job or if I rub him with my body through his pants. It's all a question of degree, and the degrees can be pretty fine – but they exist and every girl determines her own boundaries.

Before discussing the third of the three dimensions previously identified, that of self-awareness and strength of identity, there remains another form of sex tourism not located in the diagram, and that is the sex tourism of cyberspace. Reference has been made to the expositions of sex tourists on the World Wide Web. Most commentators on this have referred to the web pages of the World Sex Guide and similar pages. Pages that have 'adult' themes are popular as is shown in Table 3.1, but the listings possibly reflect patterns of access to the net as much as levels of interest by the general population. Bishop and Robinson (1998) and Kohm and Selwood (1998) are among the commentators who have noted this new form of sex tourism. For Bishop and Robinson their interest has lain in the way in which physical sex tourists report their sexual adventures and the attitudes that are being expressed. Writers engage in travelogues giving insights as to 'best places' and travellers' tips as to prices and practices. Many engage in forms of listing conquests and betraying stereotypical attitudes about women being 'great lays'. An example derived from the World Sex Guide dated 23 March 1997 describes a visit to an Auckland Massage Parlour thus:

> At the top of the stairs, an older lady in a booth gets your agency fee ($55 NZ) which gets you in the door. The lounge is a swanky but small bar environment. The girls are all pretty hot compared to escorts in other countries, as a lot of really 'respectable' girls do this for extra cash on the side after their normal day jobs. I was immediately attracted to a blonde with very short hair, great legs, and ... well, she had porn star quality breasts, at least 40D and natural (as I found out later). Her name was Lee. We started with small talk and a drink, then migrated to a theme room [i.e. theme décor]. I had no idea what was coming next, as I had never been to an open, low stress place like this before. Lee ran some water into the Jacuzzi, and invited me to get in (I did). She slipped in behind me, but not before I saw how hot her body was.

Table 3.1 Most popular Net pages – all themes (as on 13 November 1998)

Rank	*Page title*	*Hits per day*	*Total hits (since registration)*
1	CyberErotica	135,666	86,207,953
2	Ty Inc. – Home of the Beanie Babies	48,743	29,549,288
3	CyberErotica's Free Tour	35,711	20,190,142
4	Wild Rose's Sexually Explicit Amateur Home Page!	28,167	20,907,634
5	Jason's Celebrities and Supermodels	25,702	12,470,232
6	CyberErotica – Monthly Girls – Free Pix	24,361	16,363,148
7	CHATROPOLIS – Live Adult Webchat	24,190	15,498,198
8	$5 Dollar Sexgallery – XXX Quicktime Movies	19,195	13,336,093
9	CyberErotica Free Area and Amsterdam Sex Shows	16,999	11,445,993
10	In Praise of Older Women	14,091	8,552,060
11	Blow Job of the Day and Other Erotika	13,823	10,131,407
12	CyberErotica's Free Samples	12,176	7,055,453
13	18,000 FreeNudie–Picts all w/ thumb–nails	7,759	186,923
14	CARDIAC ARREST! Free Pics! Cool Stuff! Links!	7,623	6,282,709
15	Sophie's Mentertainment Online	5,982	1,1,80,595
16	Impulsive Piks On The Net	4,672	2,763,852
17	Hiroyu's Sexy World. Legs Fetish	4,546	2,454,169
18	Great Books Home Page	4,376	1,773,124
19	Ladies Online Club	3,637	1,186,106
20	Celebrity SlugFest	3,191	2,062,962

And so it continues in the vein of judging women by sexual performance – but, as noted by Bishop and Robinson (1998), also accompanied by how good the correspondent was too. And if it wasn't good for him, then the fault lay elsewhere. Rarely is there a perception that it wasn't good for the women – if the woman is considered, then it is always 'good for her' because the correspondent is 'good'.

Sex on the Internet consists of both the passive and the interactive. In terms of the passive, as shown in Table 3.1 the 'tourist' as voyeur is well catered for. It is a form of voyeurism that seemingly offers to the unwary the anonymity thought to be required as described by Ryan and Kinder (1996a). It permits apparent safety, an intimacy of privacy within which fantasy can be engaged, it is a form of relaxation, it offers like the travel brochure, the promise of being there, but without the risk of 'travel'. And, forgetting the trace left by the 'cookie' it all seems risk-free. This form of sex tourism has no risk of a sexually transmitted disease.

Of more interest as a sex tourism experience is the use of the 'chat rooms'. The anonymity is used by correspondents to be more direct and explicit than would be the case in normal society. Both men and women appear to feel removed from constraints and may be as direct as they would be in the case of their requests to sex workers. The publication of *e-mail//a.love.story//* (Fletcher, 1996) generated a number of reviews and discussions in the popular press as to the authenticity of the experience. Internet reviews generally concur that, while dated, the experiences portrayed contained considerable truth, and indeed, if anything under-stated the nature of the references to sex (for an example, see the reviews of this book on http://www.Amazon.com). As such, the work contains many examples of men using the 'net to display a greater directness in sexual interest. For example, 'Charles Leslie' writes: 'I'll tell you right off the bat, I am a married man. This medium offers the possibility of connecting with persons of the opposite sex and engaging in sexually arousing conversations that benefit both parties without risking AIDS or the emotional and marital problems which might result from adultery' (Fletcher, 1996: 11).

The female character in this book, Katherine Simmons (Kate), finds herself drawn into a series of relationships and is entranced by the power of the net. She writes:

> I can confess this to you … playing at being the temptress and the power it gives me over men is the real appeal computer relationships have for me (although, don't get me wrong, I enjoy the sexual titillation as well). I have been at the mercy of men all my life and have served them. I love turning the tables. It makes me feel invincible and young. In real life, I can't wield this power any more.
>
> (ibid.: 188)

The parallels with both sex worker and sex tourist are obvious. As already cited, Kasl describes the 'power' felt by one prostitute as she dresses and 'men come running' (1989: 154). But equally, like female sex tourists, Kate exercises and enjoys a sense of power over men as a means of feeling young. Like Wicken's 'Ravers' she takes time out from her daily reality to engage in the sense of being able to sexually want, and be sexually wanted. She wants to engage in telephone sex, she wants, eventually, to meet the men she converses with. She wishes for affection not gained from other sources. In the light of the previous discussions of motive she can be termed a sex tourist.

Thus we come to the third dimension of sex tourism, and that is the contribution to senses of self-identity. Kate's sexual life through cyberspace now begins to affect her sense of identity. She changes her hairstyle, she feels good about herself (Fletcher, 1996: 189). But finally she has to address the question of the nature of the relationship with her computer lover, is it one of infatuation, an addiction, what and who is the 'real' Kate, and where is her home in a physical and spiritual sense? Thus, the question, what do the commercial sex relationships mean to people? How important are they?

4 Bodies, identity, self-fulfilment and self-denial

A personal introduction and a caveat

This chapter is concerned with how tourists and sex workers come to terms with their respective involvements as clients and providers of sex services. As discussed by Manderson (1992) and Ryan and Martin (2001) in their respective papers, such discussions of sex work involve the authors, either as writers and/or interpreters of evidence. As mentioned in the Preface, the primary research upon which this book has been based has proven to be one where preconceptions have been challenged, and whereby individual stories represent many different experiences. Accordingly, this chapter commences with an examination of the issues researchers face when dealing with these subjects. Accordingly, the first person is used to locate and construct the text.

The immediate site of our experiences, of our perceptions, is the body we each occupy. It is a source of pleasure and of pragmatism as we adjust to its restrictions as it ages. It is the medium between the world around us and how we interpret that world. The way our bodies work shapes our perception of the world, and, as we critically appraise our bodies, the way we view ourselves. The changes of the body as we age force us to reconsider and remake ourselves. As time passes, the photographs of our youth become images of a stranger, faintly remembered perhaps, a person to whom we remain connected, but an ancestor who may, or may not have approved of our current thoughts and actions. In early youth the body is taken for granted, but with passing years it exerts itself as a source of wanting sexual gratification, and with experience, becomes a conduit for pleasure. As we grapple with the nature of that pleasure and the means by which it is obtained, so too we begin to shape our view of ourselves. If, following Simmel (1971), our certainties depend upon the existence of fixed boundaries, then the first of the boundaries that we construct in our matrices of knowledge is that of our bodily selves. But the will to know must transcend the will for certainty, and bodily and sexual awareness is about the increasing awareness of body, its sex and its relationship with a sense of self-identity. For every person a process of maturation involves a transition from a state of unknowing to caring and knowing about self as a body to developing a sense of identity which fuses body, mind and what may be termed 'spirit'. To state the obvious, sex tourism is, in part, about gratification of bodily senses, but within a specific context of, in Western Protestant Anglo-Saxon society, formal social disapproval. Whether or not full intercourse takes place, there is at least a process of cuddling, of feeling naked bodies and as is evidenced time and time again, this is important to clients (see, for example, Seabrook, 1996; Ryan and Kinder, 1996b; Plumridge *et al.*, 1997). As already noted, in this form of tourism the tourist does proceed further than simply the ocular (Urry, 1990) and in doing so must consider the presence of the body, its changes over time and its role as a source of gratification with what that means for self awareness.

Lenore Manderson begins her paper on public sex performances in Patpong with the observation:

> In draft form, this paper provoked reaction. Like the paper itself, each comment spoke of its author's politics and sensibilities, and hence made a point that is crucial to discussion of the field of discourse: writing about sex is seen as writing about oneself, and in consequence, personal judgements of morality and ethics muddy the reception of the text … writing about sex – though not necessarily sexuality – implicates the author.
>
> (Manderson, 1992: 451–452)

She goes on to say there are no safe havens, and to research and explore constructions of desire, fantasy, and lust are to render both writer and reader a voyeur. One may engage in the process of semantics if one wishes to refute the charge of voyeurism. Mulvey (1981) argues that voyeurism has arguably two components – the scopophilic instinct, that is the pleasure involved in looking at another person as an erotic object, and ego libido instincts within which self-identifications and relationships are played out. As a researcher one may deny the former, even if colleagues are disbelieving, but the latter component is more difficult to deny if possible to do so at all. It is also of interest to ask why denial of the scopophilic is thought to be necessary. In a discussion between this author and Sarah Pinwell of WISE in the ACT (Women in Sex Employment in the Australian Capital Territory) about the researcher as voyeur, the immediate reaction of Sarah as a sex worker was 'What is wrong with voyeurism; what is between a women's legs is beautiful.' To discuss the sexuality of others means at many levels seeking to make sense of it through one's own value systems, and in doing so a need arises to construct one's own framework of identity. In an interview for this book with Margaret Austin, author of *Dancing Naked: An Exhibitionist Revealed*, Austin commented to me [Chris] that one reason why I was talking to her was because of a need within myself. As was briefly discussed in the preface, it is difficult to disentangle one's role as researcher, *flâneur* and as some one engaged on journeys of discovery which require self-assessment. Ryan and Martin discuss this issue in their paper on striptease – the first from the perspective of the researcher, the second as the subject of the research. They conclude:

> The researchers can, in all honesty, only provide answers in so much as their observations have meaning for them at the time of writing. The past and current involvement of the writers as friends, and as observer and stripper, mean their view is that of insider and outsider, but above all are the views of participants. The act of participation is the source of the validity of the views expressed, and acts as the boundary which limits the meanings of those views.
>
> (Ryan and Martin, 2001)

This chapter also makes significant use of the work of Robert Stoller. As will be noted, Stoller has significantly influenced the interpretations of the motivations of sex tourists offered by O'Connell Davidson (1998) in her theorisation of sex tourist behaviour. Stoller's work is also important in providing an underlying psychoanalytical rationale for the views expressed by other writers and activists in the field, notably those associated with the Campaign Against the Trafficking of Women. However, while not agreeing entirely with O'Connell Davidson's interpretations of sex tourism, primarily on the premise that they may be too reductionist, we would agree with Stoller when he wrote that theories of human behaviour have been weakened by two convictions – first, that informants' subjectivity is a red herring in research and, second, that 'the use of the researcher's subjectivity ... must either be removed from one's methodology or have its traces erased from the record' (Stoller, 1991: 3). (For a good example of the retention of the author's subjectivity, see Abramson, 1993.) In a review of ethnographic writings Stoller continues to note that:

> There shines though the writing so little respect for the complexity of the informants ... for the complexity of the topic being studied ... for the culture at large. This defect must contribute to the thinness of the thickest ethnographies.
>
> The trouble with thickness – complexity – for me there is no end. The more informants, the more information: the more questions; the more modifications; the more exceptions; the more stories, all different. Few forests, mostly trees.
>
> (Stoller, 1991: 6)

So too with this subject of sex tourism and perhaps particularly with this subject of how participants explain their participation. I [Chris] have delayed the writing of this chapter, and perhaps regard it as the most difficult of this book. To write of sex tourism and its relationship with senses of self-identity on the part of both sex worker and sex tourist is to engage upon difficult ground. As researchers who have discussed these issues with participants and who have sought to make sense of what has been said, there is a continuing awareness that the very fact of asking questions is not a neutral act. To simply ask, 'Why do you do it?', 'What does it mean for you?', is to act as a catalyst for the very things that people may not wish to choose to ask themselves. It forces reflexivity where such reflexivity is not necessarily wanted. On the other hand, many sex workers are used to being asked such questions by their clients, and have a range of answers that are protective in nature – protective of their own selves, protective of the clients in not telling the truths that clients do not want to hear. Equally clients have a range of answers, but it seems that at times they do not know why they seek the services of the sex worker. Does this imply habituation on the part of a few, or a sense that the issue is no longer of importance? As a researcher, to get beyond ready statements is difficult and requires repeated conversations and large degrees of trust.

Lofland wrote of such a process of qualitative research thus:

> A major methodological consequence of these commitments is that the qualitative study of people *in situ* is a *process of discovery*. It is of necessity a process of learning what is happening. Since a major part of what is happening is provided by people in their own terms, one must find out about those terms rather than impose upon them a preconceived or outsider's scheme of what they are about.
>
> (1971: 4)

So, to discuss identity in this context is in itself not a neutral act for the researcher. It is a process of learning about not only the subject of the research, but the reactions of us, the researchers, to what we have learnt. It is a re-iterative process of learning and interpreting. Thus, in this chapter there is no pretence that this is a definitive statement about how sex tourists and sex workers perceive themselves. However, there needs to be made a clear statement about a view which has been reached as a result of conversations with both prostitutes and clients, and as a result of a literature review, and that is that there is significant self-denial on the part of both sex tourists and sex workers. In a discussion with Michelle McGill from the New Zealand Prostitutes Collective (NZPC), she stated:

> There is a lot of self-denial in this industry. People not recognising, not wanting to recognise what is happening. The more I see the more I am coming to believe that women are hurt by this industry. I don't want to believe it, and there are people who do succeed, who are strong, but … You know, they feel strong, and then you give them a tax form to complete, and they can't do it. They want someone else to do it; they want someone to tell them what to do. We compartmentalise – we feel in control with a client, but that is in one box, and it is not transferred into other boxes. And so women don't get out. Each day is taken by each day, there is no forward planning because that means taking control, and so many women don't seem able to do that.

This view is one that the authors share. It is therefore with this caveat that the nature of self-identity and sex tourism will be analysed. This chapter will take the form of first briefly discussing the meaning of self-image or self-identity. For our purposes these will be treated as inter-changeable terms. It will then specifically discuss O'Connell Davidson's work as being, thus far, the major contribution to a structured understanding of sex tourism *per se*, and will then pose a different viewpoint. We say different and not alternative in the sense that the work of O'Connell Davidson is not rejected entirely, but rather it is contended that it is partial whereas we seek a more holistic model. This will then be followed with abstracts from field notes and interviews published by others to describe the complexities of motive and self-image of those involved in sex work and sex tourism.

Senses of identity – theoretical frameworks

Sex tourism shares much with all tourism in being a sub-set of the entertainment industry, and like the entertainment industry, the venues are filled with actors who strut their lines in a pretence of relationships. A female sex worker is cited by Cabezas as saying:

> You have to throw yourself at them, laugh a lot, demonstrate love and affection, act as if you like them. I don't know where so much laughter and love comes from. I don't know where. You have to find it where there is none.
>
> (1999: 109)

It is a theme of this chapter that it is too simplistic to consider that such laughter emerges as if from an empty well. The well of 'laughter and love' has lean dry spells, but sometimes, perhaps to the surprise of the parties, it is a well of refreshment. But the refreshment of self is unequal, uncertain and its complexities are many, embracing self-delusion as well as genuine affection.

There are many theories as to self-awareness and its relationship with sexuality and gender. It is not the purpose of this chapter to review these theories, but a very brief outline of possible approaches will be undertaken to provide a context. From these, one specific viewpoint, that of Brierley (1984), will be used as a model for further discussion as it seems to fit the tensions that arise within sex tourism. This model is also preferred as an alternative to the theories of repression that are prevalent in the work of other writers such as O'Connell Davidson (1998). One reason for preferring Brierley's work is that it encompasses states of confusion at a conscious level that may be felt by both client and sex worker. Equally, this model will be used to suggest a means by which the 'ease about a sense of deviancy' is developed by both sex tourist and sex worker. However, the work of Stoller (1975, 1979) will be examined in some detail because it has explicitly influenced the work of other commentators on sex tourism, notably O'Connell Davidson (1998) and is consistent with views expressed by other writers like Barry (1995), Dworkin (1988) and the observations of commentators like Marcovich (1998).

It is almost now commonplace to state that while sex is associated with hormones and genitalia acquired at birth, gender is a socially determined construct. How we act as men and women is generally deemed to be a learnt behaviour, but a tension, a dissonance can exist. Thus a man may wish to dress as a woman and thus confusion arises where the behaviour is deemed to be sexual while it may be genderal. If gender is learnt behaviour, then it is associated with roles thought appropriate by society, and roles are related to senses of identity. Hampson (1986: 55) reviews the literature on sex typing and concludes that 'self-perceived sex-role stereotyping has important consequences for self-perception'. Social psychologists would also adhere to quite holistic definitions of those things that contribute to 'self'. Thus, over a hundred years ago, William James stated: 'A man's Self is the sum total of all that he can call his, his wife and his children, his ancestors and his friends, his reputation and works, his lands and horses, and yacht and bank account' ([1890] 1991: 279). Apart from the assumption that 'Self' is male, this definition is important when one considers it further. The concept of Self includes our behaviour, and our possessions, and by extension our perceptions of how others perceive us. It is an act of imagination, of emotion and how we react to others' presumed judgements.

There are, therefore, significant implications of this conceptualisation for sex tourism. If the role of prostitute is deemed to be socially unacceptable, the sex worker must cope with the image of being socially condemned. There exists a need to come to terms with a separation of self from the role, even while the role may provide economic support and networks of friends. Likewise, the sex tourist needs to rationalise his or her act of patronage in a way that is consistent with sustaining a required self-perception. This is true of any client–sex worker relationship, but being a sex tourist adds the extra nuance of what being a tourist means. The act of travel, of being away from home, of being anonymous, reinforces the imagination of fantasy, of being free from normal social conventions. To be 'sexy', to engage in fantasy, and even more to act out that fantasy, within societies subject still to the traditions of Judaeo-Christian ethics, is to engage in potentially deviant behaviour. The act of travelling – of being out of context – arguably eases the sense of deviancy. This ease about a sense of deviancy is possible for a number of reasons. First, as noted by Rock (1973), the concept of deviancy is socially determined, and yet ambiguous. Thus being a tourist during a period of socially condoned hedonism away from the normal places, roles and people that might constrain behaviour creates a sense of permissiveness about behaviour. Second, any sense of deviancy may be assuaged by arguments of the naturalness of the impulse and the values attributed to the sex worker. Third, and associated with this, they may engage upon an alternative set of ethics wherein they accord to sexual acts a means of attaining a sense of renewal. Fourth, and alternatively, sex tourists can devalue and make unimportant the act, possibly by dehumanising and commodifying the sex worker. Fifth, sex tourists can retain contradictions within their self-perception by sustaining a disintegrated self-concept. It is this last variant that will be discussed in some detail.

Rock (1973) cites Garfinkel's reconstructions of sexual typification, where Garfinkel commented:

> From the standpoint of an adult member of our society, the perceived environment of 'normally sexed persons' is populated by two sexes and only two sexes ... The members of the normal population ... the *bona fide* members ... , are essentially, originally, in the first place, always have been, and always will be, once and for all, in the final analysis, either 'male' or 'female' ... For normals, the presence in the environment of sexed objects has the feature of a 'natural matter of fact'. This naturalness carries along with it, as a constituent part of its meaning, the sense of its being right and correct; i.e., morally proper that it be that way.
>
> (Garfinkel, 1967: 122–123)

Rock comments that if this is the case this explains the attitudes towards homosexuals and towards prostitutes. The prostitute uses sexuality as a means rather than an end, and he argues 'she (*sic*) flouts the expectation that there is a clear division of labour and symbolisation between male and female roles'(Rock, 1973: 53). However, in the thirty years since Garfinkel's listing of the culture of properly sexed persons, that culture has been challenged, and in that challenge there emerge new ambiguities. If it is 'proper' that a homosexual can assert his sexuality, then the heterosexual client can claim an assertion of his masculinity by engaging with more than one partner. If the prostitute claims the role of 'priestess', the client may well be happy to claim the role of supplicant.

The nature of social change with reference to sex roles in the last thirty years has been, in the popular culture of the Protestant 'Anglo-Saxon' world of the United Kingdom, Germany and the United States quite significant. Bem (1993) has argued that in an emergent androcentric world the consequences of past notions of what are 'real men' have been particularly difficult for men. She states that it is males who are made to 'feel the most insecure about the adequacy of their gender ... crossing the gender boundary has a more negative cultural meaning for men than it has for women – which means, in turn, that male gender boundary-crossers are much more culturally stigmatised than female gender-boundary crossers' (1993: 149–150).

It has been observed by many writers on sex tourism, (e.g. Seabrook; 1996, Bishop and Robinson, 1998; O'Connell Davidson; 1998, O'Connell Davidson and Sanchez Taylor, 1999) that many clients of the bars of Patpong seem unable to cope with the changing social relationships that have occurred between men and women in recent times. The certainties once possessed by these men as to their sex role have been found wanting and unable to cope with the changes described by Bem (1993) and thus, from their viewpoint, the act of prostitute patronage is not a deviancy but a re-establishment of the 'natural order' of things. However, argue Howard and Hollander (1997), the theories of gender roles have often noted that difference hides another truth, that of inequality. They argue that gender research often tends to ignore power relationships, and they further argue that 'Stereotypes about women and men differ on the basis of the race, ethnicity, social class, sexual orientation, age, and other social locations of both the perceiver and of the target of the perception' (1997: 31). For O'Connell Davidson and Bishop and Robinson, the actions of many sex tourists in Patpong are based on racism. O'Connell Davidson writes:

> Without exception, the sex tourists we interviewed reproduced the classic racist opposition between the 'primitive', who exists in some 'state of nature', and the 'civilised', constrained by powerful legal and moral codes, in their (mis)understandings of their host cultures.
>
> (1998: 178)

Similarly, Bishop and Robinson observed that travel brochures perpetuated colonial imagery of Thai people as 'Simultaneously childlike and erotic' (1998: 89) and as 'the most sensual and overtly sexual on the earth' (ibid.: 90). Bishop and Robinson also note the racism expressed by British and American sex tourists that they are somehow 'better for the women' than Arab or Japanese clients who are crude and aggressive (for example, see Bishop and Robinson, 1998: 231), while O'Connell Davidson (1998: 180) makes similar comments. Indeed, for O'Connell Davidson and Sanchez Taylor, the racist basis for tourist demand and the need to exercise power are as equally applicable to female sex tourists as for their male counterparts. They note that 'Western female sex tourists employ fantasies of Otherness ... to obtain a sense of power and control over themselves and others as engendered, sexual beings and to affirm their own privilege as Westerners' (1999: 49).

Before turning to the work of Brierley, one further consideration needs to be examined. For some writers the sex tourist is not simply having to assuage a process of guilt arising from acts of deviancy, but is essentially and often knowingly engaging in acts of repression. The act of sex with a sex worker is a means of re-establishing an identity of self through the repression of the humanity of the sex worker. Yet it is argued that this process of objectification goes beyond the denial of emotional content; rather, emotional satisfaction is obtained through this process of commodification, and thus the act of going to a prostitute is characterised by perversion. For this reason O'Connell Davidson seeks to explain much about the behaviour of sex tourists through the theories of perversion espoused by Robert Stoller (1975, 1979).

Stoller's book, entitled *Perversion: The Erotic Form of Hatred*, has few references to prostitution, but where they exist they almost inevitably describe prostitute usage as being a perversion. For example, on page 8 Stoller wrote 'But look closely at cryptoperversions such as rape, or a preference for prostitutes'. On page 52 he cites as perverse the man 'who is impotent except with an officially degraded woman'. Stoller defines perversion as:

> the erotic form of hatred ... It is a habitual, preferred aberration necessary for one's full satisfaction, primarily motivated by hostility. By 'hostility' I mean a state in which one wishes to harm an object. The hostility in perversion takes form in a fantasy of revenge hidden in the actions that make up the perversion and serves to convert childhood trauma to adult triumph. To create the greatest excitement, the perversion must also portray itself as an act of risk-taking.
>
> (1975: 4)

That going to prostitutes, for some men, does engage a sense of risk, the risk of being found out, and this risk heightens the whole experience, is found in the literature. Often, it seems, the sense of risk is associated with an adrenalin rush of anticipation. Boyle (1994: 50) describes 'Steve' as saying, 'I get a mixture of fear and adrenalin picking a girl off the street.' He goes on to observe

> The fear of being caught never enters my head. At the same time I didn't want my wife to find out because I thought she might blame herself. I have a high sex drive but I didn't feel as if I was being unfaithful to her as there's no real emotional link with the prostitutes.
>
> (ibid.: 51)

This lack of emotionality is for Stoller, a sign of perversion. Perversion involves not only hostility, but revenge, triumph and a dehumanised object (Stoller, 1975: 9). Writing of pornography, Stoller argues that even in the mildest form of pornographic photographs, 'These reduce the actual woman to a two-dimensional, frozen creature helplessly impaled on the page, so that she cannot defend herself or strike back' (ibid.: 133). So too it is with prostitutes he argues. Prostitutes he says are 'humans hired to act like puppets' (ibid.). In his subsequent work, *Sexual Excitement: Dynamics of Erotic Life* (1979), Stoller argues that sexual arousal is based on hostility, and on needs for secrecy, mystery and risk. We need, he argues, to keep secret from others our own sexual fantasies, while the object of our desire must retain mystery – a paradox exists whereby that which is 'hidden from oneself must be made conscious and the prostitute, the lover becomes an actor in one's own theatre of the mind' (ibid. 1979: 17–18).

This approach has importance for a number of reasons. First, it moves away from the analyses advanced by writers like Dworkin (1988) and Barry (1995) whereby there is a real danger in concluding that *all* sexual contacts between males and females are simply extensions of patriarchal systems of power. This is not to say that such analyses are without value inasmuch as they permit sex workers to, in effect, argue that they are simply like many other women who are getting paid for the provision of sexual comfort – wives simply get paid in different ways. Such an argument creates a commonality of female interest. This view is clearly expressed by Nicky Adams and Nina Lopez-Jones in the interview reported by Silver (1993) in the chapter entitled 'What I do for my husband is really no different'. A common interest exists because of the economic conditions faced by women – Silver reports Lopez-Jones as saying, 'And I think you'll find that money is something that normally all women have in common when you talk about their jobs. I don't know many secretaries who would be doing it for love if there was no wage attached to it' (Silver, 1993: 100). However, the conceptualisations of an activist like Barry and the Campaign Against the Trafficking of Women do lead to real problems for sex workers as discussed by Murray (1998b). With reference to Asian sex workers working in Australia she argues that:

> The anti-trafficking campaigns actually have a detrimental effect on workers and increase discrimination as they perpetuate the stereotype of Asian workers as passive and diseased. Clients are encouraged to think of Asian workers as helpless victims who are unable to resist, so may be more likely to violate the rights of these workers. The campaigns also encourage racism towards Asian workers within the industry … and in the general community where Asian workers form an ostracized new 'underclass' without equal rights.
>
> (1998b: 58)

Stoller's analysis moves us from the general argument of women entrapped in a global penile colony (to misappropriate Robinson's term – see Robinson, 1993) to the specific nature of the contract between client and customer. To contract for sexual services, to use what is often an unequal economic power, particularly when dealing with sex tourism in the developing world, is a very specific example of sexual relationships, albeit one related to wider social gender relationships. As Høigård and Finstad carefully note:

> It is important to maintain the connection between men's 'usual' sexuality and men's experiences with prostitution. Prostitution magnifies and clarifies some of the characteristics of masculine sexuality … But if the end result is exclusively a catalogue of the entire panorama of motives for male sexuality, then the structuring of the question and the methodology are wrong. If one primarily wishes to chart men's usual sexuality, then prostitution research is a detour and skewed point of departure. The approach to the problem should probably be different. Why do some men seek out prostitutes in order to act out their sexuality? What qualities does prostitution have that make some men specifically choose prostitutes?
>
> (1992: 93)

O'Connell Davidson uses Stoller's theses because she argues that the Barry and Dworkin approach, by ignoring the nature of the 'free contract' between sex worker and client, ignores an important component of prostitution. She argues that:

> I believe that the façade of voluntarism makes prostitution all the more insidiously damaging to many of the women, children and men who work as prostitutes. What I *am* arguing is that, if you overlook the way in which a fiction of consent is constructed, if you dismiss the fact that, as a rule, both parties to the 'exchange' buy into this fiction, and ignore the tacit rules which govern this kind of interaction, you end up telling a story about prostitution which is unrecognisable to most of the people who actually participate in it, and thus cannot hope to *explain* why they do it.
> (O'Connell Davidson, 1998: 121–122 – original author's emphasis)

Others have questioned the nature of the 'contract' between client and sex worker. Thus Høigård and Finstad wrote:

> There is another complication in defining prostitution as a voluntary activity … In order for the woman to trade in her sexuality in the marketplace, she must treat it as an object that can be relinquished and made use of as the possession of a stranger. She must be prepared to separate sexuality from its position as a part of her own identity, her own personality. She must have learnt to split herself into an object and a subject. Her own sexuality must be an object that she can manipulate and transfer.
> (1992: 180)

In the world of popular entertainment the nature of the contract between sex worker and client as a 'victimless crime' has also been exposed as being, at best, an incomplete truth. In an episode of the TV series *Cracker*, the playwright Jimmy McGovern puts into the mouth of a disillusioned wife the words that 'all that time he was spending forty quid on her [a prostitute] while I was having to scrimp and save'. In short, money spent by clients, who are married, on sex workers, is money denied to the family budget, and in a sense therefore there exists a further confirmation of the (economic) subordination of women. Thus the argument is that prostitute usage by men is about issues of revenge, fetishism, objectification and commodification of women. It is a perversion. It is 'an act of oppression' (O'Connell Davidson, 1998: 121).

Within this thesis of perversion lay subtle nuances and problems that are discussed by Stoller. Stoller's work is Freudian in tone – trauma or frustration in childhood is suggested as a possible cause of the erotic hatred of women. But he also proposes a notion that sexual excitement is aroused 'when adult reality resembles the childhood trauma or frustration' (Stoller, 1975: 105). The orgasm within the perverted state is 'not merely discharge or even ejaculation but a joyous, megalomanic burst of freedom from anxiety' (ibid.: 107). The problem about this approach, this attribution of hidden, unconscious forces, is to reconcile it with what actors in this particular drama are saying about their experiences. Stoller cites a prostitute talking about her work after being a sex worker for a year. She says:

> Part of the excitement was seeing a guy's genitals or feeling them. But I just don't get the same reaction now as I used to when it was more or less a mystery to me … you see all these men; and at first you are really excited about them being men. That in itself turns you on and gives you orgasms … But now, to achieve an orgasm is like having to almost battle for it … A lot of times, I get more excited when I have my clothes on and he has his clothes on and we are making out and playing around. I get really excited then.
>
> (ibid.: 107–108)

Stoller reproduces a longer section than this, but the process of familiarisation is made clear in this shortened extract. However, Stoller begins his analysis of the text by stating that inevitable curiosity about sexual differences arises in small children, and further comments that a major way for the looking to be sexually exciting is for a man to believe he is acting forcefully, sadistically upon an unwilling woman. This implies that for Stoller, the prostitute's initial excitement makes sense only if she is viewed as a small child, while his comments about the male view do not appear to follow from the text of the prostitute's comments. The problem about developing a psychoanalytical theory of male needs to dominate, and of a perversion based on such needs, is that it seems to ignore many contrary statements being made by the respondents. O'Connell Davidson arguably displays the same tendency. For example, she comments after discussing one case, (that of 'Guy'), that his 'sexual excitement hinges on the hostility and vengeance involved in reducing a human being to those bodily parts' (1998: 154). At another point she comments that 'all the clients discussed thus far are clearly sexually hostile men' (ibid.: 143). However, the text provided offers alternative readings. Guy requires 'mothering' he converts female sex workers 'into goddesses', he requires antidotes to loneliness – the reading of his motives presents a state of confusion and ambiguity within which the role of sexual hostility is not overt, and is but one possible strand of the mental state of the client. Perhaps the main problem that we have with this mode of interpretation is that it ill fits the data raised in our own researches, some of which is reproduced below. Another problem is that, as a theory of sexual motivation by sex tourists, it is primarily addressing heterosexual male motivations and thus ignores the fact that there are growing numbers of female, gay and lesbian sex tourists, and nor does it differentiate between the various modes of sex tourism demonstrated by the figures in Chapter 3 of this book.

Before leaving the psychoanalytical theories of Stoller it also needs to be said that those who adhere to his theories, or similar theories, do not use the full range of conceptualisation enunciated by Stoller. The largest individual case study that he discusses to illustrate how 'risk and revenge are formed into excitement in perversion' (Stoller, 1975: 123) actually relates to a discussion of a male homosexual sex worker. Stoller argues that his sexual excitement comes from the passing of money into his hands and comments that 'excitement, stimulated by the sight of money, comes from hostility' (ibid.: 124). Thus Stoller's theories of perversion are not directed solely at explaining male attitudes towards females, but at a homosexual's hostility towards 'straight society' (ibid.: 199–203), and arguably could be extended to female hostility to males. There is a selective use of these theories and furthermore an avoidance of those things not consistent with a thesis of male hostility to females as an explanation of sex tourism. O'Connell Davidson thus sets aside the issues posed by female sex tourists as being a minor thing on the grounds that 'heterosexual female sex tourism is, in numerical terms, a far, far smaller phenomenon than male sex tourism' (1998: 181). (It is of interest that subsequently, in 1999 she treats female sex tourism more fully – see O'Connell Davidson and Sanchez Taylor, 1999.) Likewise she rejects the views expressed by a writer like Bell (1994) as representing a discourse of a 'handful of relatively privileged white American women' (O'Connell Davidson, 1998: 113), and that this discourse should not be accorded equal weighting with more typical samples of prostitutes. A further problem with the adoption of theories of male hostility towards female sex workers is that it generates, of necessity, a poor perception of the sex worker herself. If her function is to act as a conduit for male hostility, what does it say of a woman who permits herself to fulfil this role? At the very best she can only be perceived as a victim, and possibly as, at best, a stupid victim in permitting this continued abuse. It denies the possibility of the sex worker not being other than a victim, and by implication argues that women who do otherwise are simply misguided, and are unknowing victims. Too rigid an adherence to psychological theories like those of Stoller also contains the danger that the respondent is not listened to.

The client's sense of identity

Stoller's work also creates an uneven picture of the client. The picture of the client that emerges from these analyses is of men unable to cope with the changing roles of women. They are men seeking a return to a patriarchal system at best, and at worst, are men consumed with (suppressed) hostility towards women. There are male behaviours that provide ample data for such views. Stark and Flitcraft (1996) provide ample evidence of the violence done to women within domestic situations, while there is equally ample evidence of violence towards sex workers. Records are compiled by prostitutes collectives whereby warnings are distributed among workers as to 'ugly mugs'. Scambler (1997) and Faugier and Sargeant (1997) provide evidence of the violence which characterises the lives of sex workers. Yet, if the models posed in Chapter 3 have validity, such behaviours by men are but part of a much wider range of male behaviours towards sex workers. Anecdotal evidence collected within the process of research for this book included the not uncommon view that if one wished to see who the clients of sex workers were, then the next ten men one met when walking down the street would provide as good a sample as any. In short, sex workers maintain their clientele are drawn from a wide range of the community. To adhere to the thesis that male clients are hostile to women and inclined to violence born out of an inability to cope with changing social conditions thus implies that all men are like this if it is accepted that male clients are drawn from all parts of society. There are very few structured studies of male clientele. One study by Kruhse-MountBurton (1996) was of 301 clients. She concludes that:

> Clients of call girl prostitution ... were in many ways typical of the male population. Single men and divorcees were over-represented, indicting that the institution provides an important outlet for men without partners ... The men varied enormously in physical presentation, ideal weight and general fitness, but their average score is a reflection of the relative youth (mean age was 32) of the sample. They were better educated and more likely to be self-employed or in management than the average Australian male, indicating that the cost of the service is a consideration.
>
> (1996: 113)

In terms of socio-demographics, apart from a bias towards higher income and education, there was nothing specific to demarcate her client group. In terms of emotional state, 51 per cent were categorised as being positive, 4 per cent were demanding and aggressive and three individuals were feeling suicidal. The remainder confessed to varying degrees of emotional concern. Indeed, 40 per cent expressed some anxiety about their relationships with women, but these ran through a continuum of concerns from worry about being alienated from their wives to wishing to escape from the constraints of marriage via upset over the loss of partners. Our case is not that there is no evidence to support the thesis of male hostility towards women, but that there is evidence for complementary and simultaneously different expressions of male thought and behaviour towards sex workers and women. Thus any theories of male and female identity need to recognise these factors. In short, we would agree with Kruhse-MountBurton when she writes: 'The normality of the client profile, and of their behaviour within the commercial encounter, is indicative of the inadequacy of explanations which insisted that the primary role of prostitution is to cater to the aberrant and psychologically maladjusted' (1996: 114). These conclusions are supported by other literature. For example Høigård and Finstad noted that:

> When the researchers compared the men with and without prostitution experiences, the most striking factor was the *similarity* between the two groups. Customers can be found in all age groups, parts of the country, occupations and social strata. Among the customers there were slightly more single men, and there was slightly more dissatisfaction with cohabitation and sex life among those who were married. Also customers were

somewhat more frequently men who travel a lot and men with money. But the tendencies are not very pronounced.

(1992: 28)

If, then, we are to establish a theoretical construction that does have general applicability, such a theory needs to be able to encompass ambiguity, uncertainty and self-denial at one end of the spectrum, and at the other be able to conceive of a full and conscious recognition of perfectly knowing who one is and what one's behaviour is – either as sex worker or client, and furthermore that theoretical construct needs to be non-gender specific as both men and women are clients of sex workers. For example, Boyle (1994: 41) cites 'Kevin' as saying, 'I am scared of being caught. One of my worst fears is what my family would think if I was found out ... I still have feelings of guilt. I was brought up a strict Methodist.' Kevin must thus reconcile his practice of visiting prostitutes with his family role, the guilt he feels, and arguably can only do this by maintaining a dissonance within his own self-image. Similarly Kruhse-MountBurton notes that of her sample six clients declared their 'animosity' towards prostitution as an institution and she writes: 'These negative opinions demonstrate that client behaviour often diverges from the clients' stated views' (1996: 113). However, the views she attributes to these specific clients are generally moralistic, which thus again raises the view that some men engage in an action for which they feel a need even while having doubts as to its moral value.

Possibly the very generality of this modelling may render it of limited usefulness, but if it permits a development of the specific then it may be of use. That ambiguity of self-perception lies at the heart of sex tourism is very clearly shown by the work of Kleiber and Wilke (1995). German tourists who were travelling to a country, which they had previously visited and had contact with a prostitute, were asked if they intended contacting a sex worker again on their travels. About 60 per cent said that they were. They were then asked if they would consider themselves as being sex tourists. Of the 60 per cent only a fifth indicated that they considered themselves to be sex tourists. The data also showed that 19.1 per cent of the respondents would, 'most certainly' revisit the previous sex partner, and a further third would 'probably' or 'perhaps' do so. About 40 per cent would either 'probably not' or 'definitely not'. Such data raise questions about definitions, but more significantly it raises questions about motivations and self-perceptions. However, in suggesting a general theory a caveat has to be added, and that is that the conceptualisation suggested here is intended to be applicable to primarily the Anglo-Saxon world. Cultural differences are important. For example Hobson and Heung (1998) report that it is not uncommon for Hong Kong males to have mistresses and concubines in Shenzhen in Southern China. They cite a Chinese saying, 'people ridicule poverty but not prostitutes' (ibid.: 137). Yiu (1994), from an interview of 200 Hong Kong residents, found that 54 per cent of Hong Kong males would take a concubine if the opportunity arose. On the other hand, 56 per cent of the wives would worry if that were to happen and 41 per cent would seek to divorce their husband. Twenty-nine per cent of the males stated they knew of people who did have a concubine in Southern China. The need for duplicity on the part of males is seen by such statistics. Knowing duplicity on the part of such males does not, however, immediately support theories of male hostility to women as discussed above, but at the very least it does require males to develop means of living with that fact of their behaviour. It requires them to incorporate within their self-perception the knowledge of their ability to sustain 'omissions of the truth' if not specific lies.

Brierley (1984) argues that the principal models of sexual deviance are five-fold, and these he lists as:

1 The classificatory model: this is often based on a link of symptoms, which linkage he argues is often tenuous. Within classification arise subsequent sub-divisions. He provides the example of transsexuality – does the person stand or sit to urinate? Such classifications, as already noted by Howard and Hollander (1997), may be the source of inequality of power. In discussion with the first author, Michelle McGill (NZPC) observed that the sex industry is one of division and sub-division where the oppressed seek in turn to oppress. Thus she noted that within the transsexual community the 'true' transsexual is one who sits to pee, thereby implying further crisis of identity for the transsexual who does not.
2 The psychodynamic model: this is the approach typified by the work of Stoller (1975) and it conceives of sexual variation as perversion. They tend to be Freudian in approach, and therefore, as noted in the case of Stoller's work, often associated with traumatic happenings or relationships with parents. McGuire *et al.* (1965) is cited by Brierley as another researcher from this school. McGuire *et al.* suggest that patterns of sexual behaviour are shaped around fantasies and objects by concomitant erotic reinforcement in masturbation.
3 The biological model: essentially this approach seeks explanations based upon physiological and genetic influences. From this perspective it might be argued that genetically males are polygamous and thus their patronage of prostitutes is explicable by this simple 'fact' alone.
4 The sociological model: here, sexual diversity is like any statistical distribution, and thus within a society a range of sexual behaviours may exist. However, society has to then respond by classifying some forms of sexual behaviour as desirable or otherwise – thus the difference in practices (or different explanations) observed by anthropologists like Malinowski (1922) or Mead (1977).
5 The human rights model: this again, argues Brierley (1984: 68), is a model based on social interaction, but here the issue is not solely that of 'desirability' but the rights of minorities to engage in various behaviours, and the means by which society develops and maintains its tolerance.

Brierley (1984: 69) rejects all of these approaches and contends 'that it is essential to look at the total personality organisation of the individual to understand his or her sexual behaviour'. It was, at the beginning of this chapter, argued that self-perception was an act of imagination by which we saw ourselves through our own eyes and through how we imagined others perceive us. Brierley adopts a similar position by arguing that the personal construct system needs a series of elements such as 'me as I really am', 'me as I wish I was', and 'me as others would see the real me' (Figure 4.1). Following the humanistic school of psychology as represented by the works of Rogers (1951), Maslow (1970) or Hampden (1971), it follows that the well-integrated personality would possess high degrees of congruence between these elements, whereas the dissonant personality has an internal complex set of relationships which create insecurity through an inability to be confident about his or her world. The model also makes a distinction between gender role and gender identity as described previously. Gender role is not a self-concept *per se*, but a persistent pattern of behaviour arising from the perceived expectations of others who surround the individual in question. Gender identity, on the other hand, is a core component of self-concept, and as such is generally firmly established. Potential for dissonance can exist in a number of ways. First, there may be a gap between how an individual thinks others perceive him or her, and how they do, in fact, so picture that person and his or her gender role. Thus it may be rendered more difficult for individuals to interpret behaviours displayed towards them. Second, if an individual is experiencing conflict over gender identity, as arguably is not uncommon with the clients of prostitutes, dissonance may occur due to the ambiguity within that core component of self-identity.

The sex object in this model is the person, or perhaps object or action, on whom or which sexual arousal is centred in the broad sense. The optimal relationship between sex object, gender role and gender identity might be thought to be one of equilibrium and to contribute to a state of homeostasis within this model. This is slightly problematical to our minds. In Brierley's (1984: 70) conceptualisation he notes: 'It will be increasingly difficult to determine an adequate sex object the more disparate the set of self concepts becomes.' Adopting psychoanalytical approaches it may be argued that the more confused the self and gender identity, the more it will respond to stereotyping. Thus the 'whore' represents a specific sex object and the dissonance is projected upon the sex worker – this would be consistent with the work of Stoller. It is also consistent with a Jungian analysis. As Samuels comments:

> Many of the men who came to see me for therapy were manifestly and magnificently confused about their gender identity ... Now what I noticed was that, for these profound feelings of gender confusion to exist, there had to be an equally profound feeling of gender certainty to be in operation at some level. You cannot know the detail of your confusion without having inkling about the certitude against which you are measuring your confusion.
>
> 
> (1995: 10)

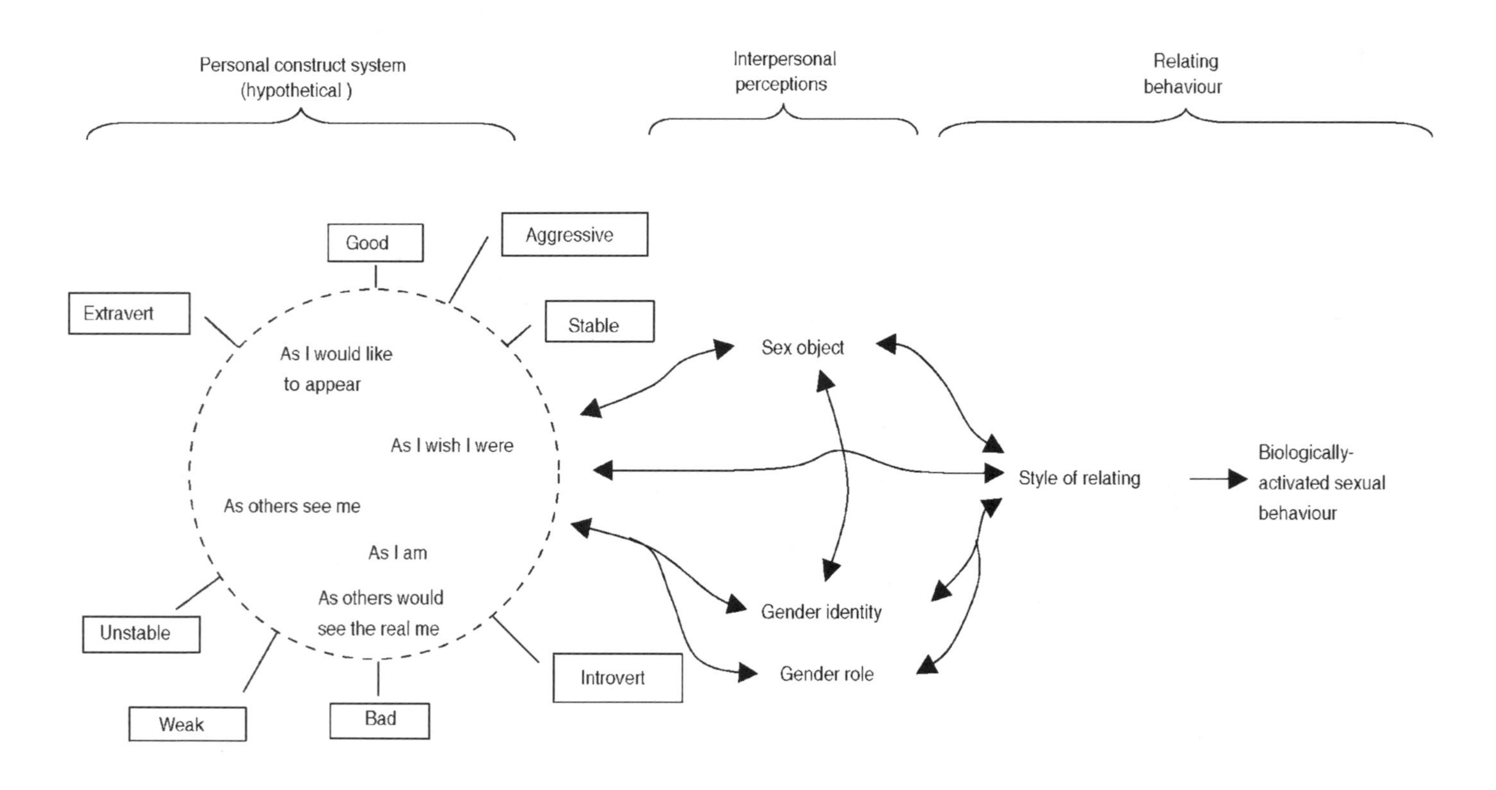

Figure 4.1 Brierley's model
Source: After Brierley, 1984.

Moreover, the 'healthy' integrated personality may also seek to disturb the process of homeostasis referred to by Brierley. For Maslow his conceptualisation of the self-actualised personality is one who is:

> relatively spontaneous in behaviour and far more spontaneous than that in their inner life, thoughts, impulses etc. Their behaviour is marked by simplicity and naturalness, and by lack of artificiality or straining for effect ... His unconventionality is not superficial but essential or internal.
>
> (1970: 157)

This spontaneity, according to Hampden (1971), is characterised by an ability to take psychological risk and it can be postulated that the self-actualised is able to sustain a non-homostatic state if required in processes of maturation of self as a state of transition. This state of transition needs, somehow, to be conceptualised if, at least from the perspective offered by Bell (1994) and her descriptions of sex workers, positive states of well-being associated with incongruities of role and identity are to be incorporated into a general model. In short, positive states of well-being can be associated with incongruities of role and self-identity as a form of ludic involvement.

Brierley's model is of interest in enunciating the concept of tension within the personal construct system and he suggests, as seen in Figure 4.1, a number of dichotomies such as bad vs. good, unstable vs. stable, weak vs. aggressive. Such dichotomies are easily found within the statements of both clients and sex workers. This is most clearly shown in the statements of men who seem to generate a process of denial that the relationship that they engage in is a commercial transaction. This is found in the work of many commentators, and is common to studies of both sex tourists and those who patronise sex workers within their own home areas or countries. For example, Plumridge *et al.* (1997: 174–177) reproduce a number of comments made by clients of New Zealand prostitutes such as: 'I feel it's more a friendship, lover type of thing than a sex worker ... It's closeness, it's warmth, it's caring ... We are friends and we are lovers' ... 'a lovely dual feeling of loving, caring or something'.

Günther (1998) describes a sex tourist's rationale of why he was not a sex tourist by reference to friendship, companionship and affection, and by the lack of payment of a fee. In notes made from interviews by these authors, one Australian tourist spoke of trips to Indonesia in this way:

> I will not engage in anything that a woman does not like, but the intimacy of sexual relationships which have been satisfying mean that a state of affectionate relations arise ... you have a laugh, you joke, it's as if you've chatted up a nice girl and you get on well. So, you go out like boyfriend and girlfriend and a sense of sharing things develops.

However, and it seems to be a sharper focus for sex tourists travelling overseas than for those who patronise prostitutes in their home country, the evidence of different realities needs to be faced. Both Seabrook (1996) and O'Connell Davidson (1998) make this clear in their work. Seabrook describes the experiences of Vince:

> I realised she had dumped me. I was badly hurt. I loved her …
>
> I know now that's the way it is. They're doing a job. And their job is to get what they can. I don't hate her. I feel hurt. I got into something deeper than I knew. I thought I could handle everything, I was in charge of my life.
>
> (1996: 35)

Vince thus has to face the reality of the commercial transaction. If there is a difference between sex tourism in Western locations, and the sex tourism of locations in South-east Asia, it possibly lies in the veil which is drawn over the monetary value of the transaction. Yet sex workers in First World societies have their regulars, and both parties will talk of friendship patterns and gift buying. There is little empirical evidence on this issue, but it can be postulated that men who perform the role of regular clients may also wish to deny the basic commercial nature of the transaction.

In explaining male attitudes and the reconciliation between 'me as I am', 'me as others see me', 'me as I would wish to be' and the other components of self-image, it is possible to draw upon other aspects of Turner's theories of the liminoid as described in the earlier chapters. As noted, Turner utilises concepts of performance and ritual. The use of a performance role as allowing one to step into an other than normal situation and act according to the requirements of that role, while distancing yourself from your usual self has been observed by sex workers and researchers. Writing of striptease, Goffman (1967: 269) notes that the striptease is a public performance, 'it is the final mingling of fantasy and action' with scenes for hire where 'the customer can be the star performer'. Patrons, it is argued, buy the thrill of 'jeopardy during a passing moment' (ibid.: 268). Ryan and Martin (2001) comment that being at the striptease is, for the man, the act of possession in the sense of being able to gaze upon the naked female form, but without the risk of performance that is involved in visiting a prostitute. It is a mimetic jeopardy – it mimes the desire but without its final fulfilment.

Schechner says of ritualistic events like striptease that:

> A person sees the event: he sees himself seeing the event; he sees himself seeing others who are seeing the event and who, maybe, see themselves seeing the event. Thus there is the performance, the performers, the spectators; and the spectator of the spectators; and the self-seeing self that can be performer or spectator or spectator of spectators.
>
> (1982: 297)

So one reason why the client participates in striptease is to be seen by other males. But why is he lured in this way? One reason may be to play the role of 'knowing man' among male colleagues. The act of placing the money in the stripper's garter is to be the catalyst for further 'action' for the pleasure of him and his mates. That pleasure may be in enticing the state of undress, or in now engaging the dancer in conversation. Through an evening, if the client stays any length of time, there is much naked female flesh to be seen, and one can view without participating. The extra that the public purchase gains is seen by all beside the purchaser. What is not shared with the wider public is the opportunity for 'private' smiles, whispered conversations, the possibility of holding the woman's discarded clothing – in other words, the purchased interaction is to go beyond the simple view of naked female flesh. It is the fantasy of possession of an attractive female and the reality of fleeting moments of communication denied to others. Ryan and Martin (2001) argue of striptease that: 'it is complex theatre where men pursue the view of the dancing naked female form, appreciate, at least in many cases, the art within the dance, and play not the role of predator but preyed upon. It is an erotic game – but a game with rules'. This concept of the performance role may have some application to visits to prostitutes if the visit becomes a means by which fantasy is enacted. The client becomes able to act out the private fantasy of success with other men or women, to achieve the comfort of being able to make contact with flesh or to assuage the feelings of loneliness (Kruhse-MountBurton, 1996). Whereas striptease has a mimetic jeopardy, according to Goffman (1967), the relationship with the sex worker can go beyond the mimetic. It is of interest that Kruhse-MountBurton reports that only 25 per cent of her sample of clients suggested that 'intercourse was the sole requirement of the transaction (1996: 123), while 10 per cent claimed that intercourse was not all that important. Yet, nonetheless she still concludes that the mystification of the commercial association permits men to interpret their involvement with sex workers 'in terms of the ideal masculine script' (ibid.: 126). This is much the same conclusion reached by Plumridge *et al.* (1997). They conclude that men exhibit no crass enjoyment of 'force and coercion' (ibid.: 178). They argue that clients relate to commercial sex as another form of the discharge of obligations, in that non-commercial sex requires other forms of payment. Yet they argue that the male rhetoric of mutuality is associated with another rhetoric, that of freedom of obligation. Male clients seek to obtain a pleasurable experience, and part of that pleasure is the thought that their commercial partner also obtains pleasure. Yet the men do not wish for a sustained relationship, and they rarely consider the implications of the sex worker who needs to assess her next client even as they (the client) depart. However, such views are not necessarily based upon gender, but upon the function. Female clients of male sex workers also provide evidence of the same functional patterns of thought. As already noted, Wickens has written of the enjoyment of lust by female 'ravers' and the very explicit modes of behaviour of such females – such as the strategy of saying 'Look at what mum has bought me' (showing a packet of condoms to her potential sexual partner)' (1997: 157). Thus, Pruitt and LaFont (1995: 425) note that Western women can construct a new feminine subjectivity in terms of the significant other, 'the boyfriend'. They seek to distinguish this form of tourism from sex tourism by labelling it as 'romance tourism' but the relationships they describe are very similar to those described by Günther (1998). Pruitt and LaFont also describe the males as being used by the women, as being the 'brown baby' to demonstrate liberal views, and thus men might resort to violence to maintain dominance. Just as Bishop and Robinson (1998) and O'Connell Davidson (1998) argue that the way men in Thailand and Cuba relate to their commercial partners, and are related to by those self-same partners on the basis of stereotyping, so too it appears Jamaican men and their white female clients inter-relate with each other in the same way.

It would thus seem that, regardless of gender, clients engage upon personal discourses of self-denial as to what their relationship means for the commercial partner by not being able to, or not wishing to, empathise with the situation of that sexual partner. It is possible to explain this in a number of ways. One is that the client's self-denial is reinforced by modes of behaviour and self denial engaged in by the commercial partner, while another is that the act of sex with a commercial partner is separated from concepts of self by investing it with a sense of ritual which reinforces the sense of performance of a role rather than an engagement of self. For the sex tourist this is made all the easier by locating the transaction in other than the normal place of residence. The rhetoric of 'freedom of obligation' becomes associated with a role that places the person beyond their normal place. Hence the self-placement into a liminal position helps protect the positive self-image of the client by the act of separating and distancing the act of client-hood from the major part of his or her life. The motives for clients are, for the most part, what clients say they are, a search for warmth, cuddles, enjoyable sex, and they live with or deny the commercial veil by the deliberate act of distancing, separation and making liminal their client-hood.

Turner (1974) also refers to the role of ritual in the process of liminoid states. Kruhse-MountBurton (1996: 149–174) identifies eight rituals of commercial sex based on her experiences as a sex worker. These include 'man-making' rituals where the sexual initiation is undertaken by a visit to a sex worker and the conclusion of the sexual act becomes an affirmation of masculinity either in a solitary manner or by the man now being able to boast of his prowess with fellow males. Male bonding is the desired outcome in some instances. Another ritual is the 'buck's night' prior to marriage. This of course has its parallel with the 'hen's night', and both are associated with sexual imagery. On one occasion the first author was having a meal with friends (including other tourism academics) in a restaurant in Rotorua, New Zealand, when the restaurant became filled with women, and the bride to be was bedecked with a veil covered with condoms and pictures of male models. Rituals of consolation and submission might also be applicable to the situation of the sex tourist. From our perspective the important aspect of this ritualistic analysis is that they are consistent with Turner's (1974) thesis that liminality places men (and women as clients) 'outside the total system and its conflicts' with the result that 'transiently they become men (and women) apart' (1974: 241). By adoption of being 'apart' they engage in a paradox whereby the confirmation of their sexuality, which is an important component of a sense of self, is achieved by re-locating themselves in a ritualistic space of the 'other'. This act of 'separateness' is aided further by travel to a place away from home. The ritual may also be encouraged by the procedures and layout of the brothel, or the pattern of approach to the street worker. As already noted, the red light district is a place apart. On entering a brothel a place of subdued lighting may be entered. Introductions are made, a drink may be served, conversation is entered into. The client is absorbed into a place separate from his or her normal existence. In the case of the beach boy, the client is approached and chatted up. The beach boy will have a routine which may involve offering to take the tourist to different, untouristy places. This separation from normal place and action may contain within it the seeds of dissatisfaction and disillusionment exactly because the need for confirmation of self is bound in a spatial or temporal limbo. It remains restricted in place and or time. Or it may be confirmatory of sexual self-image in the way that all ritual can reconfirm, even though it is a 'time out'. In this latter case the client is able to transfer the experience to their normal milieu. The statements of clients recorded by researchers like Kruhse-MountBurton (1996), Plumridge *et al.* (1997) and in research by the authors would indicate that this special time can, but not necessarily, have positive results in some cases. However, the liminality of the experience contains a potential and continuous weakness for client self-identity because as Høigård and Finstad state:

> The prostitution customer shares the generally accepted goals of the male role: objectification and subordination of women. But the customer uses normatively illegitimate means to obtain this goal: instead of women's 'voluntary' subordination he resorts to money's tyranny.
>
> (1992: 105)

To this might be added the fact that the female client equally purchases sexual favour rather than being able to depend upon 'female allure'.

There is no doubt, however, that for many clients the commercial veil is non-existent – it is a simple commercial transaction that by its nature characterises the freedom from obligation. Ryan and Kinder (1996a: 512) cite 'Ted' who had no time for emotional involvement, saying 'I don't want to have to go through all that bullshit'. For such respondents the motivation is one of wanting a sexual release.

The sex worker's sense of identity

The concept of the sex worker as occupying the space of an 'other' has been fully discussed already in the preceding chapters. However, with reference to sex tourism, it is perhaps worth re-iterating that this appears to be a fairly general concept, and examples can also be found within the Third World. Thus Laga (1993) describes occupants of *lokalisasi* (licensed brothels) as being women from outside of the region, even though no statistics are available, and furthermore the reason for concern is because

> [the] infiltration of prostitution is considered as a threat to the culture and, as a result, to the profile of Nusa Tenggara Timur woman which will cause it to become weak (*perinfiltrasi prostitusi dianggap sebagai terhadap kebudayaan sehingga profil wanita berkepribadian di Nusa Tenggara Timur menjada lemah*).
>
> (Laga, 1993: 5)

Sex workers are thus marginalised as a threat to the social fabric. Any actions which reinforce marginalisation will have an impact upon the psychological well-being of the sex worker. Ryan *et al.* (1998) describe how, in the city of Christchurch, New Zealand, police require sex workers to register with them, and in this process of registration, a photograph is taken of the sex worker. This is how this one sex worker, Claudia, spoke of her feelings about working in Christchurch.

> I came here to New Zealand about four months ago with my boyfriend and about NZ$30,000, but we split up and because of that one mistake I have lost $15,000, and I have to earn it before I can go back to Germany or to see a friend in the USA. But what can I do – I tried to get other work but it is impossible – in Germany I worked as a receptionist in a hospital but here – it is my language – I am not so good in English. So I have to do this – I do not like it, but what can I do? I smoke to help me do it, but I don't take other drugs – at first I cried and I didn't like it and I didn't get orgasms. I don't like men, but if you go with a woman friend they say 'Lesbian' you know. Any time I get involved with a man it gets me down. I have no husband, I have no children, I don't like this, but I need the money. If I work in a shop, what would I get, $500 a week, but after tax that's about $400, but I can't live on that as I have to pay rent, for food and so on so I work as this.
>
> You know the police come in here – they come in and they come right into the rooms. But I haven't done anything wrong, I have a visa – I can understand if I have been speeding, or have a parking fine that I appear on a police record, but what we do in here is private and they have no right – but my name is on their computer, but I am doing this for only a few months to earn money so that I can leave – so why should I be on their computer – what will happen in the future? One day I had to go to the police station about something different, and in there I saw the policeman who had been in the parlour, and he saw me – so what will they do – they call me prostitute, so will they help me?

This extract from research notes clearly identifies a number of common themes within prostitution – problems with men, the need for money, and additionally a sense of stigmatisation by society which means that they are denied access to many of the social support structures that many citizens would take for granted. To sustain, under such circumstances, any sense of psychological well-being would require considerable confidence and a strong sense of self. Yet there is little doubt that some women do achieve this, while more women would seek to claim such strength. However, when we have recourse to Brierley's model as shown in Figure 4.1, sex workers have greater problems in sustaining this positive sense of well-being within the culture of Anglo-Saxon societies with a Judaeo-Christian tradition. Male clients, as discussed, can have recourse to a sense of confirmation of their masculinity and within male bonding a sense of behaviour condoned if not condemned. Female clients may reconcile themselves with thoughts of establishing equality with male behaviours. Female sex workers do not have such easy consolations. They are perceived as either the victims of male hegemonies, as betrayers of their sex through the explicit commodification of their sexuality, or as trophies of men that reinforce psychologically weak men in their conceptualisation of masculinity. They may be seen as brazen hussies or as psychologically disturbed themselves. Stoller, writing of porn stars, describes such women as possessing a hysterical personality and he writes that:

> Organic or psychological theories can account for such [erotic] intensity. A brain abnormality – for example, an anatomically well-placed tumour or electrode – can spark fierce erotic behaviour, though I doubt that it would produce the raging hysteria seen in these women unless the patient was already a hysteric.
>
> (1991: 24)

Thus the female sex worker is characterised as deviant, abnormal, or as a victim of men, of drug taking – in short, any number of factors exist that can reinforce low self-esteem. However, many of these attributions are projections upon the sex worker who becomes a fantasised object – a fantasy born not solely of the erotic desires of actual and potential customers, but also of the different agendas of researchers and social guardians. Their normality is oppressed by these writings and their role as women seeking alternatives to low-paid occupations is minimalised. Many are women who may be without formal educational qualifications or who may find it difficult to re-enter the workforce after periods of child bearing and looking after young children because employers do not impute value to home keeping functions. In many instances women find themselves in child keeping roles after divorce or separation, and thereby having to face limited paid employment opportunities. Thus many women enter a socially despised occupation from a position of economic weakness, a factor which further compounds difficulties in maintaining a healthy sense of psychological well-being. These negative factors may be further reinforced by women who seek to escape dysfunctional family backgrounds. A significant literature exists which describes this; for example, see Caplan (1984) and Gibson-Ainyette *et al.* (1988).

Consequently, in terms of the model derived from Brierley (1984) and shown in Figure 4.1, the social imagery which relates to 'how others see me' is associated with several negative images. The sex worker must thus seek to overcome those feelings, which are internalised. As shown by the above extract from Claudia, work in the sex industry may be associated with tears and drug taking in order to cope with the fact that one is the plaything of clients. One of the 'discoveries' of the new sex worker is that most men are actually quite 'nice', many are nervous, many are submissive, and many are complimentary. Many also come very quickly and the experienced sex worker knows many ways of making a man come quickly so as to avoid penetration if she does not wish it. The literature of relating to prostitution and the sex industry is thus replete with many statements of sex worker satisfaction about the easy money, about the dressing up, and a sense of power over males as has been earlier used in this book. In their work on strippers in Darwin, Ryan and Martin (2001) cite the example of Jasmine, who began work at the age of 30 and felt very positively about herself as a sexual being. In research for this book another exotic dancer, Roxanne, spoke about conflicting senses of well-being thus:

> I started stripping 24 weeks ago. Why? I've been in debt a long time and just didn't want to continue and so had to earn some money. Also I had my second child about 6 months ago – I was depressed and had low esteem – I was critical about myself, about my figure, about my breasts which are important to men. I had to earn money. I like dancing and this was a job I could do with dancing.
>
> I have a much higher self-esteem now, what have I learnt? Not to be afraid of men. My husband knows, but he hasn't seen me here, but he wants to – he is still working up on the Sunshine Coast but is hoping to get a job down here. I haven't told my children, haven't told my friends – no-one knows – I have had to create a new circle of friends here – but the women here are great – they taught me what to do – and if one is being successful I see what they do and follow that – I like talking to people, but you have to give all the time – I need affection – you don't get affection here. It's giving all the time. My parents don't know – they are very religious – I was too – they are Seventh Day Adventist, but the church is in a mess – it's falling apart. Yes – the body can be a source of pleasure – I dance for my husband.

Again this extract is replete with economic need, questions of the speaker's sexual appeal which is important to her, the sense of newly acquired confidence, processes of revaluation, of cognitive dissonance, and of having to make adjustments to a new life-style. In one sense exotic dancers have a more difficult regime than prostitutes in having to be confined to clubs for long hours and having to work long shifts. However, these features are shared by those who work in massage parlours, where, as noted, systems of fines may be in place. Independent prostitutes who work alone or in the company of other women in sex worker co-operatives do not have to remain in the one location for many hours, especially with the advent of the cellular telephone. Modern technology allows them to shop, care for children, laze on a beach if they wish, talk to a researcher and still conduct business.

In discussing this issue of self-reappraisal with one New Zealand sex worker, Bobbie, a response was provided that illustrates the compartmentalisation of self that can occur. Bobbie said:

BOBBIE Before I started in this game I didn't pay much attention to astrology, but I think there might be something in it. I am a Gemini, you know, 'the twins', and I have two personalities. There's 'Bobbie' – she's a fun-loving, extrovert, having a good time person, and there's another me, the old grump, and Bobbie wouldn't like her.

AUTHOR But the 'old grump' is the one who is 'Mum' and worries about the bills and providing for the children. And I suppose that 'Mum' had within her personality the embryo of 'Bobbie'.

BOBBIE Yes, so I suppose there is three personalities. 'Bobbie', 'Mum' and a mixture of the two. You know, my children know that I'm sometimes called Bobbie and if I get called that in my normal life that's not a problem, but if I get called my birth name in my job – it's like [makes a gesture of her jaw dropping], it's a

AUTHOR intrusion?

BOBBIE Yes, it's not supposed to be there.

One factor that aids this process of re-evaluation is the realisation that one is not alone. As Roxanne comments, 'the women here are great – they taught me what to do'. This solidarity of the sisterhood of sex workers is again a common feature of the writing of sex workers themselves. Sisters of the Heart (1997: 22) speaks of 'compassion and camaraderie amongst the girls and management. It feels like one big slumber party or a bunch of girls living together at a college dorm.' On one occasion this author was speaking about the boredom of the women hanging around in the massage parlours waiting for clients when Michelle responded – 'No, it's not always like that – sometimes you get a client and you get rid of him quickly because we girls are enjoying ourselves too much.' More formally, Hanson (1994, 1996a, 1996b) has commented on the way in which women form supportive circles for each other. Hanson (1996a) notes that 'examples of peer-education, mentoring and support systems are readily found in the sex-industry'. Kempadoo and Doezema (1998) provide several accounts of more formal support mechanisms based, generally, on prostitute collectives around the world, some of which are funded by various health councils or health education agencies. These formal and informal arrangements are important in creating a peer group with which the sex worker can relate. Brothels also have their social life apart from the sex workers and the clients. They are the basis of a mini-industry where, for example, seamstresses, some of whom many be former sex workers, will come in to sell costumes. A paradox exists whereby expensive clothes which are quite skimpy, but which require many hours of work because of the sewing on of sequins by hand, are sold, and of course, often discarded quite quickly by the wearer in the course of business. Ryan and Martin (2001) comment on how quickly exotic dancers get through pairs of shoes. Other sellers enter the doors of the massage parlours selling items that may have fallen off the back of the proverbial lorry. And, in some cases drugs may be sold and consumed, although this is not overly common in licensed brothels as many legislatures would withdraw the licence if drugs were found on the premises.

One side-effect of this milieu is to normalise the working environment for the sex worker. The processes of familiarisation and habituation are important in creating a sense of the normal. Within the normal the sex worker may be able to develop a platform from which she can state her occupation in a wider social circle. For example Momocco, a Japanese sex worker, writes that she 'appreciates her work' and tries to provide a luxurious service and takes pleasure from those occasions when a customer says 'thank you'. 'But I still need courage to speak out like this. I always try to tell my friends about my work, but I don't want my parents to find out' (Momocco, 1998: 180). For Momocco the process of commoditisation of sexual pleasure legitimises her work. The growth of the leisure industry has meant that pleasure has become a commodity, and sexual pleasure is thus too a commodity. As a provider of sexual pleasure she is a worker, and thus sex work should be an established form of labour like any other.

Hence for many sex workers their positive self-image is generated by the normalisation of their work and the environments within which they find themselves. 'It is a job, like any other job' is the view propounded. Homeostasis of self-identity within our model is maintained by an equation of self as worker and by the sexual other becoming not an object of erotic fantasy, but simply a client, a source of income. While there exists a potential issue of the client becoming de-personalised, the commercial reality is that this is avoided in the same way as in the case of any other service situation. Utilising the dimensions of the SERVQUAL model (Parasuraman *et al.* 1988), namely 'tangibles', 'reliability', 'responsiveness', empathy' and 'assurance', the brothel seeks to perform 'excellent service' like some bank. By commodifying it normalises, and by commodification the sex worker assumes a professional pride. It inserts codes of professionalism between the emotional trauma and economic truths of inequality of power. Sex work becomes but work, like any other. Behind this approach lies a functionalist rationalisation of sex work, but it possesses for its adherents the advantages of removing stigmas of victimisation. One such example of this is provided by Athina Tsoulis (1999), in her screenplay for *I'll Make You Happy*. In it the boyfriend says to his sex worker girl friend, 'I guess you must be quite experienced.' 'Oh no,' she replies, 'I'm keeping myself for someone special.'

In many ways this proposition of commodification is a managerialist view, and represents a simplification in terms of how individual sex workers may feel about their situation – yet it may aptly describe one trend. However, for writers like Bell (1994), Sprinkle and Gates (1997) and Kruhse-MountBurton (1996) the positive self-image is based on factors beyond commodification and normalisation. They are among those who claim that sex work is about naturance and the rediscovery of the spiritual meanings of male–female union. Veronica Vera maintains that:

> Sex is a nourishing, life giving force and as a consequence sex work is of benefit to humanity ... Sex workers are providing a very valuable service to be honoured. Sex work ... is a good service, it is the best service that one individual can do for another individual.
>
> (cited by Bell, 1994: 108)

For Kruhse-MountBurton masculinity is an evolving experience for men, a relational concept which stands with and complementary to the feminine. She notes 'Men hold most of the power in society and yet they are the sexual and emotional supplicants who depend upon women's bodies for nurture and reproduction' (1996: 14). Her sex worker role is thus one of the nurturance of masculinity, not as a sex object to be used, but as a sexual being to develop masculinity. Bell writes that the prostitute is to claim back the position of the priestess of ancient days, as 'worker, healer, sexual surrogate, teacher, therapist, educator, sexual minority and political activist' (Bell, 1994: 103).

For these women, a sense of positive well-being is associated with a sense of spirituality that may be said to have some association with New Age philosophies. It is thus difficult, as we have seen, for some writers like O'Connell Davidson, to accept this position, especially when they consider the position of Third World sex workers and the conditions that attend the lives of many such women. The present authors are unable to speak with any direct experience of the conditions of the lives of Thai or Filipino prostitutes, not being able to converse easily in their languages, but it can be observed that there are different voices and complexities. Montgomery (1998) notes the war of words that have broken out between women who come from the sex workers' perspective, and those who, without direct experience of sex work, wish to speak for them, perhaps for reasons of their own agendas. In our own research, as has been argued, we have not been able to sustain the thesis of male exploitation as a sole explanation, finding instead other voices. Yet, equally, a continuous refrain does exist that for many women prostitution is a result of constrained economic choice, even within the Western world. And those structures have been primarily determined by past and current power structures dominated by men. We are also in a difficult position when it comes to assessing the position of child prostitutes in Third World countries, but colleagues well known to us have undertaken research and through our knowledge of them we respect the integrity of their findings. Cooper and Hanson (1998) report talking to sex workers in Vietnam who, they assess, were aged about 13 or 14 years old. They concluded that, by treating sex work as work, and by facing the realities that economic options are severely constrained, it was possible for Vietnam through safe sex programmes and sex worker education to develop a local sex industry based on the principle of worker control.

Hanson (1998) also questions the rationale of campaigns aiming at the abolition of child sex tourism on the premise that the real issue is not child prostitution, but the capitalistic system that gives rise to the poverty of those people whose young daughters have to enter sex work. She argues that the liberal position of condemning child prostitution is itself a denial of the power structures imposed by global economic systems. Again re-iterating her theme that sex work is but work, she argues that the liberal position, given the realities of capitalism, and the culture of familial support structures in societies such as Thailand, actually means a denial of income to poor families. Better, she argues, to improve working conditions for sex workers. Hanson is not alone. Montgomery (1998) also argues that child prostitutes had very different self-constructed social frameworks than those attributed to them by Non-Governmental Organisation campaigns. She notes that in these campaigns the children themselves are rarely allowed to speak – they are given the role of passive victims whose role is to await rescue. Her own research in Thailand found an ethical system whereby children possessed strategies for rationalising their participation and for taking pride in their support of family through their earnings. Yet the rationalisation was in terms of friendship. The money veil referred to above was a means which helped to protect the children's own sense of worth. To be called a *sopheni dek* (child prostitute) or a *ying borikan* (business woman – a slang expression for a prostitute) was a hurtful insult. Thus the paradox emerges whereby the 'money veil' which is the basis of O'Connell Davidson's (1998) and Bishop and Robinson's (1998) criticism of sex tourists as not facing the truths of their actions is also the same mechanism which protects the prostitute of the Third World and enables them to retain their sense of integrity. The 'fact' is recognised, but the evaluations differ, and the 'real' fact is the subjective evaluation made of the objective!

Brierley's model is of use in identifying the social relationships that impinge on self-identity. It reinforces the link between the external and the internal and in doing so draws our attention to the role of culture and milieu that are so important in determining perceptions of self and evaluations of behaviours. Combining this with the socially and economically determined marginality of different groups within society permits a means of analysis which moves away from the simplistic and reductionist positions based on moralistic approaches. It also reminds us of how, in this subject of sex, so much of our understanding is still based on a Judaeo-Christian tradition by which the Anglo-Saxon world has eroticised the Asian world without adding to our understanding and has assigned to homosexuals a past position wherein they perhaps occupied a place of greater condemnation than that occupied by sex workers. It is these latter worlds that will be considered in more detail.

Finally, how does this relate to the previous chapter? There it was argued that sex tourism could be re-classified on the dimensions of commerciality and exploitation. A third dimension is the retention of integrity, and the power structures at micro- and macro- level that can either threaten or support the sex worker's integrity. Hence the diagrams of Chapter 3 need an added dimension. This is shown in Figure 4.2. Thus an additional dimension of a retention/confirmation of self-integrity at one end of the continuum leading to a denial or suppression of that integrity at the other end of the continuum is added to the previous diagram. Sex slavery through exploitation and degradation for commercial ends denies any sense of fulfilled self-identity to the victim. Yet even this representation portrays the complex and ambiguous nature of prostitution and sex tourism. Given the nature of the previous discussion about sex tourism and self identity of both clients and sex workers it becomes obvious that both may occupy more than one position on this dimension. It does, however, help us understand the pluralistic nature of sex tourism. Jordan, writing of the sex industry, noted that:

> The search for 'representativeness' itself constitutes a false scientific objective. It ignores the fact that each of these accounts [of her sample of prostitutes] is valuable in its own right, and enhances our knowledge and understanding of women's involvement in sex work.
>
> (1991: 6)

As one of our sex industry workers who has helped us by reading this script said:

> I don't think anyone in the Prostitutes Rights Movements really expects that the majority of workers love their job so much that they would not change it if they had unlimited opportunities. What we DO think is that at this place and this time, these workers are doing this job because it is the best choice we can make. Like any other workers, there are some of us who like it more than others within that group. Also some of us pride ourselves on a job well done and there are some of us who really aren't concerned, as long as the guy doesn't complain. We would be doing sex workers a disservice to categorically state that all or even most sex workers are mad keen vocational hookers. My experience is that most of us are seasonal or situational workers. We work to pay off the car, send the kids to private school, get through university, etc. Some of us will go on to be career prostitutes, but most of us will step out of the industry with no long-lasting emotional scars from our experiences.

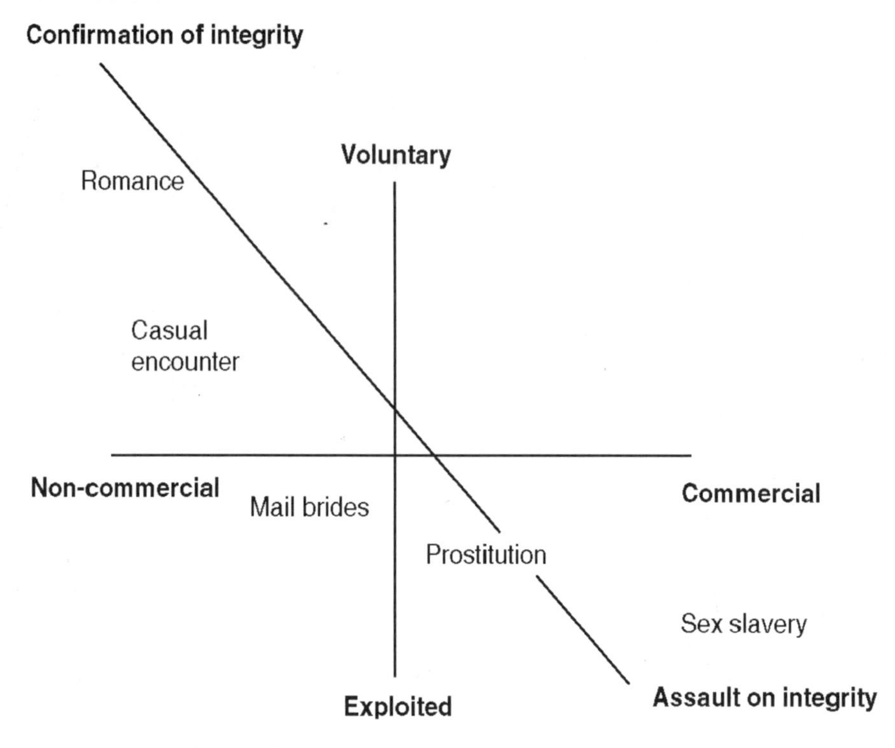

Figure 4.2 Paradigms of sex tourism

Whether the client is a local person or a tourist, or whether the sex worker is male, female, heterosexual, gay, lesbian or transsexual, if there is a simple truth about this issue, for those women not in conditions of slavery, then this statement might be one that could attract a consensus among sex workers. But there are many separate voices. To paraphrase Oscar Wilde from *Lady Windermere's Fan*, some may be in the heavens, many are in the gutter looking up at the stars, some enjoy being in the gutter, and some are bound to the gutter by the force of others' dysfunctioning or that of their own – they are all voices to be found in the sex industry, just as it is in life more generally.

5 Bodies on the margin

Gay and lesbian tourism

The idea of the sex tourist and sex tourism occupying liminal space has been central to the development of this book. The notion of Turner (1974) that liminal entities are neither here or there, that they are betwixt and between the positions assigned and arrayed by law, custom, convention and ceremony has been utilised to help partly explain the cultural space that sex tourism fills. As has been discussed, the tourist is therefore a marginal person, with the extent to which tourism is sanctioned varying from society to society and the nature of such marginality changes over time. For example, Lea (1993: 709) noted that in the early 1960s 'the advent of young westerners and their casual attitudes to sex and drugs had a severe impact on the predominantly Roman Catholic Christian Goan community'. Writing several years later Wilson (1997: 70) noted that there is 'little evidence of hostility and resentment among local people against the majority of such tourists', and argued that this is because, in part, such low-budget, low-income tourism in Goa has spread the economic benefits of tourism more widely through the local community than five-star hotel development. Wilson quotes an interview by Anderson with one villager in north Goa who stated:

> The villagers … were devout Catholics and they had been offended by hippies bathing naked in front of their homes. But that had now stopped. [They were] not bothered by the hippies' lifestyle. The village as a whole were happy to augment their incomes by renting out rooms in their houses to budget travellers and running little cafés and bars which fell well short of overwhelming village life.
>
> (1997: 28)

In 2000 Goa actively promotes itself as a tourism destination to the West, often with reference to the image established by the 'hippie' period and is now a major location for the hosting of rave dance parties on the beach.

A range of lifestyles and activities may occupy this liminal space; Beddoe (1998) mentioned it in association with beach boy related tourism in Sri Lanka. Similarly, Lett (1983) in investigating the sexual behaviour of individuals on yacht charter in the British Virgin Islands, found that the customary courting relationships were abandoned during the holiday period and individuals were more apt to engage in sexual relationships with strangers. Based on Turner's (1974) work, Lett conceptualised 'liminoid' as having liminal qualities without a ritual component (Currie, 1997). According to Lett (1983: 45), 'Liminoid activities, in short, are those socially accepted and approved activities which seem to deny or ignore the legitimacy of the institutionalised statuses, roles, norms, values, and rules of everyday, 'ordinary' life.' In Lett's study, charter tourists were engaged in a liminoid experience, with an altered point of reference, which allowed then to behave almost opposite from their normal behaviours in their home environment, including their sexual behaviour. Nevertheless, as Currie (1997: 892) noted, 'Given the opportunities and resources of a destination or the particular needs of individuals, this sexual inversion may not be possible or desirable for individuals.'

Graburn argued that such reversals or inversions of behaviour are a necessary part of life and imply that, 'certain meanings and rules of 'ordinary behaviour' are changed, held in abeyance or even reversed' (1983a: 24). Graburn depicted inversions as polar opposites that reside on a continuum that allows individuals to determine the degree to which behaviours are reversed or inverted from their lifestyles in their home environment, with each kind of tourism being 'characterised by the selection of only a few key reversals' (ibid.: 21). Wagner (1977), in reference to male prostitution in the Gambia for the predominantly female Scandinavian market, argued that the 'inversion' brought about by such relationships, was perverting the norms ruling gender relations in many societies. Wagner argued that there was a destructive potential in these relationships:

> What to the tourist is a pleasant and refreshing interlude where the disregarding of norms in no way threatens the structures pervading in their home society, could result in the destruction of one of the very foundations of local social structure, that of ordering social life according to age and generation differences.
>
> (1977: 50)

Nevertheless, such inversions are not a given, as Graburn noted:

> reversals are bi-directional, that a number of the polarities are related to each other; that rarely are tourists motivated by only one kind of behaviour reversal; and that tourists only choose to switch a few of their behavioural parameters at any one time, while retaining the vast majority of normal repertories.
>
> (1983a: 24)

If, as the Preface to this book has argued, sex tourism has occupied a marginal position in tourism research, so issues of homosexual roles are even more marginalised. Indeed, the position of women, ethnic minorities and homosexuals in tourism is almost as marginalised in the tourism academy as it is in the articulation and representation of heritage, identity and tourism (e.g., Edensor and Kothari, 1994; Kinnaird and Hall, 1994). Nevertheless, in recent years a number of significant attempts have been made to redress the imbalance in studies of gay tourism (e.g. Hughes, 1997; Pritchard *et al.*, 1998). In this chapter the expression 'gay', generally understood as the historically specific US slang term from the 1970s, now often internationalised, travelling globally via gay politics, gay culture and gay tourism is used in preference to 'homosexual' or 'homosexualities', the attempt at a universal, pluralistic (but uneasily, is it only male?) term, which deliberately situates and re-situates the medico-legal discourses of the late nineteenth century to the present day (King 1994).

In the year 2000 gay tourism as an expression of gay lifestyle is marked by stereotypes, homophobia, significant misunderstandings and, often deliberate misrepresentations. Right-wing church and religious groups in particular often seek to portray homosexuality as some sort of crime against God and typically seek to deny the gay community the rights which are assigned to most individuals in society. In particular, the expressions of gay pride at community events and festivals that often represent, to some, portrayals of indecent and lewd behaviour, are often represented as the 'norm' to the mainstream community and 'dangerous' to society (Daniels, 1999). Indeed, certain homophobic groups often equate the homosexual community with the actions of paedophiles in terms of being sexual predators or engaging in 'public orgies' (IRN News, 1999). Such representations clearly match certain portrayals of sex tourism in terms of its potential to 'undermine' or 'affect' society. However, as in the case of sex tourism there is a multiple layering of meaning in the expression of gay lifestyle through travel and community celebrations, as there is in any group in society. Indeed, we recognise that there is a danger in sectioning out in a book on sex tourism a specific discussion on gay travel as it may reinforce the representation that the expression of gay love is somehow deviant. That is not our intention at all. Instead we want to illustrate the marginal space of liminality through a range of different settings and how they may be read and interpreted. Indeed, in examining aspects of tourism in relation to gay and lesbian travel we also want to highlight the manner in which the tensions of marginality may actually contribute to greater understanding and tolerance in society.

One of the great myths of gay lifestyles represented in some of the mass media is that it is full of single gays and lesbians who are seeking casual sex. As with all myths it is based on truth. However, within an examination of homosexual lifestyles there is no reason to believe that there is more casual sex than in the expression of heterosexual behaviours. For example, Clift and Forrest (1999) in a study of gay travel motivations, noted that 'opportunities for sex' ranked substantially lower than many other motivations.

Despite the stereotyped image of homosexuals as people seeking loveless sex it is not unusual for gays to be in monogamous relationships. For example, a

Canadian report indicates that most gays and lesbians would rather be in a stable relationship or married and are not just interested in casual sex. A poll of 1,700 gays and lesbians for *Fugues*, a Canadian gay magazine, indicated that 55 per cent of the respondents were in a stable couple and 69 per cent wanted to get married. According to Michael Hendricks of Operation couple, a group that wants to demystify the social and sexual lives of gays:

> It breaks the stereotype that we are lonely men going around looking for other lonely men … We buy candy and flowers for each other at this time of the year [Valentine's Day] and everyone that's alone is hoping that Cupid will strike.
>
> (cited in Canadian Press, 1999)

Rejean Barbeau, who was helping launch Operation couple, said, 'People have the impression there's no stability in the gay community when there is … But that impression distorts the reality. There are couples who are together as well as singles' (cited in Canadian Press, 1999).

While many myths abound regarding gay lifestyles, it can be noted that travel and holidays constitute an important part of gay identity in Western society. As Hughes (1997: 6) observed, 'Tourism and being gay are inextricably linked. Because of social disapproval of homosexuality many gay men are forced to find gay space … Gay space is limited … and gays find it necessary to travel in order to enter that space.' The extent to which gays travel has made them a lucrative travel market (e.g., Holcomb and Luongo, 1996; Clift and Forrest, 1999). According to the 1999 report on the gay and lesbian travel industry by Community Marketing (1999) the gay and lesbian travel sector is worth approximately US$47.3 billion, or about 10 per cent of the US travel industry total. Gays and lesbians have a higher propensity to travel than the non-gay population. According to the report, 85 per cent of gays and lesbians surveyed took a vacation in the previous twelve months, compared to a 64 per cent national average. Thirty-six per cent took three or more vacations, while 45 per cent went overseas, compared to just 9 per cent of the national average. Gay and lesbian income demographics reported in this travel survey are similar to those of other studies (e.g., Pritchard *et al.*, 1998). Many gays and lesbians have a higher household income level. Some 75 per cent have incomes of more than $40,000 per annum, and 23 per cent have an income over $100,000 per annum (compared to 9 per cent of mainstream American travellers). Though one of the most significant indicators of the gay markets propensity to travel was the result that of the survey population 78 per cent held a valid passport, compared to just 29 per cent of the mainstream. Similarly, in Australia, *Campaign* magazine's research shows that gay men earn and spend much more than the population average. The study also revealed that companies who had marketed themselves toward the gay spending dollar were already reaping large profits. Conducted by Significant Others Marketing Consultants, the survey results have gay men earning significantly above the national average with almost 40 per cent travelling overseas the previous year, more than half eating in a restaurant at least once a week and 23.6 per cent dining out two or more times a week. The survey also noted that more than half would change banks if offered a gay credit card (Brother Sister, 1996a).

In the American survey sea cruises were particularly attractive to the gay and lesbian travel population with the survey finding that 15 per cent had taken a cruise in the previous twelve months. By comparison, only about 2 per cent of the mainstream population cruised in the same period. In this market gays and lesbians were choosing mainstream though 'gay friendly' cruise options rather than exclusive tours. To date, only Carnival and Windstar cruise lines have openly sought the gay and lesbian market with advertising and promotions. In terms of airlines, the survey reported that among domestic carriers, American Airlines was rated 'favourite' for the fifth year in a row, primarily because of the airline's visibility in the gay community, sponsoring charities and gay travel-related promotions. In fact, according to Community Marketing's research results, 'giving back' to the gay community – in the form of donations and other support – was an important factor for 89 per cent of survey respondents when choosing among travel supplier and agent options (Community Marketing, 1999).

Given the degree of homophobia in certain sections of society, 'many gays will choose to travel in search of an anonymous or safe environment in which to be gay' (Hughes, 1997: 5). Therefore, perhaps unsurprisingly, being a gay friendly location acts as a significant factor in travel planning. Among American urban destinations, San Francisco rates first, with 49 per cent having visited in the past three years. Resorts such as Palm Springs and Key West also remained significant destinations. Activity- and event-related trips are also very popular among the gay community. In the previous three years, 50 per cent went to a gay pride event, 13 per cent travelled to attend a circuit party, and 4 per cent participated in a 'gay ski week' event (Community Marketing, 1999).

Where destinations are overtly unfriendly to gays, then their tourism industry may suffer. For example, in light of a number of concerns raised by gay and lesbian tourist interests, the Jamaica Forum of Lesbians, All-sexuals and Gays (J-FLAG) called on government and tourism interests to make Jamaica more appealing to the international gay and lesbian tourist market with the organisation being concerned that Jamaica's international image of not being welcoming to gays and lesbians will damage the country's already fragile tourist industry. J-FLAG (1999) notes that Jamaica's internationally known homophobia will alienate it from markets in countries that have no sanctions against sexual practices between consenting adults. They observe that Jamaica may also come under increasing pressure from international gay and lesbian organisations, such as the International Gay and Lesbian Travel Association (IGLTA), whose Executive Director Augustin Merlo on 30 December 1998 stated that his organisation strongly condemns Jamaican authorities:

> for failing to ensure the safety, welfare and comfort of gays and lesbians living on and visiting the island nation. The island nation appears to be a

> leader in the region's emerging homophobia that has already shown its ugly face in the Cayman Islands and Costa Rica. That is a tragedy ... If Jamaica is unwilling or unprepared to welcome gay and lesbian tourists to [its] shores, then IGLTA is prepared to warn all of our member companies and associations that our tourist dollars are no longer welcome in that country.
>
> (quoted in J-FLAG, 1999)

J-FLAG believe that significant fallout in the tourist market could result from the IGLTA statement, noting that if Jamaica continues to be seen as a destination that discriminates against gay and lesbian tourists, it will lose out on the multi-billion dollar international gay and lesbian tourist market, which South Africa, with its non-discriminatory policies, has aggressively begun to court. Therefore, J-FLAG was calling on the Jamaican government and tourism interests to put in place measures aimed at welcoming all tourists regardless of their sexual orientation, including repealing the nation's buggery law and a change in the interpretation of the gross indecency law.

Despite the positive tourism image of Australia in the gay travel community because of the success of the Sydney Gay and Lesbian Mardi Gras, Australia has a long history of homophobic attitudes and laws, many of which are still in place (Brother Sister, 1996c). In the mid-1990s, the State of Tasmania in Australia introduced a series of anti-gay laws (since overturned), which Tasmanian gay activists claim were costing the State valuable tourist dollars. For example, at a tourism conference in Hobart, the state capital, the General Manager of a tourism promotions company, Landmark South Pacific, Judy Ashton, said that Tasmania was missing out on the lucrative gay travel market because of the perception overseas that it was illegal for two men to book a room together. Tasmanian Gay and Lesbian Rights Group spokesperson, Rodney Croome, also confirmed that visiting gay couples reported being denied accommodation on the basis that homosexual activity was still illegal in Tasmania. According to Croome:

> If we are to have any chance of tapping into the gay and lesbian travel market our anti-gay laws must either be completely repealed by the Parliament or conclusively invalidated by the High Court ... For this reason we call on those members of the Upper House who oppose gay law reform to put their personal prejudices to one side in the interest of job creation.
>
> (quoted in Brother Sister, 1996b)

In contrast to the Tasmanian situation, the State Government of New South Wales began to openly support gay tourism in 1995. Opening the 1995 NSW Tourism Conference, which for the first time included a workshop on the niche market of gay and lesbian travel, State Premier Bob Carr cited the Gay and Lesbian Mardi Gras Festival as an important factor in boosting Sydney's international profile as a tourist destination and announced that Sydney had been voted the world's best city for tourists in a survey conducted by the magazine publishers, Condé Nast. In contrast to the previous conservative state government which ordered the state tourism organisation to remove reference to the Mardi Gras in its promotion, Premier Carr made specific mention of the significance of the Mardi Gras Festival to the state's tourism industry, not only in dollar terms but also its celebration of arts and culture. In addition, Lynne Hocking, founder of the travel group Destination DownUnder whose company organises holidays for more than 1,500 gays and lesbians a year, speaking at the conference, said that mainstream travel agents were increasingly educating their staff to meet the needs of gay and lesbian travellers in a bid to be regarded as gay friendly, concluding that professionalism without prejudice was the key to being part of this growing market (Hopkins, 1995). Further signs of change in the Australian mainstream tourism industry with respect to gay travel was the decision in 1995 of the national tourism promotion organisation, the Australian Tourism Commission (ATC) to part-fund a brochure aimed specifically at the gay and lesbian market, representing the first time that the ATC has targeted the gay niche market. The brochure was produced with the state tourism offices of Victoria, NSW and Queensland, after the ATC pledged to match other contributions dollar for dollar (Brother Sister, 1995).

Sydney's Gay and Lesbian Mardi Gras has a substantial spin-off effect across Australia. For example, international visitors to Mardi Gras spent an average of $347.25 per day and stayed for about twenty-one days, outstripping general tourist spend by 30 per cent. In relation to the effects of the 25th Mardi Gras held in 1999 the corporate manager at F.O.D. (Friends of Dorothy) Travel, Greg Miller, stated:

> The US market is always huge, but this year they have really strong buying power ... One of the things people forget is that these tourists come back every year. There is very strong repeat business and when they're here, they spend ... Nine out of 10 Americans include Cairns in their trip.
>
> (cited in Pettafor, 1999)

Nevertheless, despite such financial success gay tourists are not always welcome. For example, in February 1999 the Australian Gay and Lesbian Tourism Association hit out at claims by Opposition Tourism spokesman Graham Healy that holiday resorts marketing gay and lesbian holidays in Queensland after the Sydney Mardi Gras was not in the best interests of Queensland. Mr Healy said that he would not like to see Queensland holiday resorts targeting only certain sections of the community and noted that he is only concerned that Noosa's up-market French Quarter resort may be tagged with a reputation for attracting one type of customer. Thus he stated:

> It would, I think, be unreasonable to suggest that the whole area surrounding that particular resort could be classed as something that it certainly is not ... In Queensland we have a reputation for attracting tourists from all walks of life, and certainly national and international tourists, and I

> think that to specifically say that a particular area could become a particular individualistic area for a particular market is something that I think is unreasonable and not in the best interests of tourism in Queensland.
>
> (quoted in ABC Newslink, 1999)

Opposition to gay tourism and events is also seen across the Tasman in New Zealand. The Hero Parade and associated festival, New Zealand's equivalent of the Mardi Gras has been held every year since 1991, except 2000, when funding problems caused the cancellation of the parade though other events went on. Described as 'Auckland's favourite free event' (*Express*, 1999), Hero was set up in 1991 as a pride event for gay men and has developed into a two-week gay and lesbian festival, which incorporates the parade and the party and a host of other cultural events. Its mission was to raise awareness and money for HIV/AIDS organisations, which at the time received very little funding. Hero Project Director Steve Berry-Smith described Hero as an umbrella organisation for the gay community.

> It's an umbrella organisation that provides a forum for all kinds of organisations and individuals to express themselves ... For a lot of people, particularly in small towns, coming out can be very difficult. The Hero event can give them something to identify with, a beacon if you like.
>
> (*Express*, 1999)

Indeed, Hero attracts participants from all over New Zealand while gays from overseas also participate on their way to or from Sydney's Mardi Gras celebrations. In 1999 publicity regarding Hero's poor funding situation assisted in raising its profile in the corporate sector and increasing sponsorship. According to Berry-Smith:

> Interest and awareness are definitely up from the business side and the public. It's one of the biggest events in the country now and Hero is pretty much a household name. We've received increased support from large corporate sponsors this year, which indicates how the acceptance level for Hero has grown.
>
> (*Express*, 1999)

However, not all public comment was positive. The Christian Heritage Party called for a boycott of sponsors of the Hero Parade, saying the procession is nothing more than a 'public orgy'. Christian Heritage Party leader, Graham Capill said that it was 'sickening in the extreme to see companies sell their corporate souls' to attract more custom. Graham Capill wanted 'every decent New Zealander to send a message to Qantas by flying Air New Zealand, to Pepsi – by buying Coke, and to *Metro* – by reading *North and South*' (IRN News, 1999).

More overt homophobia was an advert placed in the *New Zealand Herald* of 12 February 1999, featuring photographs of Dame Whina Cooper (a Maori rights activist), Mahatma Gandhi, Mother Teresa and Martin Luther King junior, stating it 'takes more than a parade to make a hero'. Readers who endorsed 'traditional family values enough to oppose the promotion of destructive sexuality' were invited to post a coupon to a group called Stop Promoting Homosexuality International (NZ). The Rev. Bruce Patrick of the Auckland Baptist Tabernacle Church said the group, which was newly established in New Zealand, was trying to raise money to help pay for the advertisement from a 'silent majority': 'You get the feeling that it is very, very hard to present any view that isn't "politically correct" through the media, that the media has a party line, and views that don't fit with that are very difficult to present' (Daniels, 1999).

Complaints about the advertisement were lodged immediately with the New Zealand Human Rights Commission and the Advertising Standards Complaints Board. Kevin Hague, executive director of the Aids Foundation, said the advertisement was offensive:

> I wonder if it had been about Maori people or Jewish people, or one of the minority groups that those four people pictured had belonged to ... There's no mention of the churches they are involved with, which is very deliberate, and no mention of anyone who is actually involved with this organisation.
>
> (Daniels, 1999)

Chief Human Rights Commissioner Pamela Jefferies said that the advertisement was unfortunate noting that the New Zealand Human Rights Act protected a wide range of different groups from unlawful discrimination, including gay and lesbian people: 'Attempts to stir up ill feeling against any of those groups are destructive. Such behaviour is inconsistent with the spirit of a tolerant and inclusive society' (Daniels, 1999). As events turned out, the 1999 Hero Parade, opened by the Prime Minister, attracted the biggest-ever spectator crowd in its nine-year history with about 200,000 people flocking to the event to see more than fifty floats moved down Ponsonby Road, featuring everything from a chorus of male Marilyn Monroe lookalikes to gay garden displays (TVNZ, 1999).

The Sydney Mardi Gras: from lock-up to frock-up

> Sure, you think you know Australia. Sydney. Big bustling capital. Opera House. The Pacific's gay Mecca. Home of the world's biggest annual gay bash, the Sydney Gay and Lesbian Mardi Gras. Or maybe it's the wild and dusty Outback, made famous by road-tripping drag queens or baby-snatching dingoes. Or even the country's famous north-eastern gold coast, decorated geographically with its own version of a geographical feather boa, the amazing Great Barrier Reef.
>
> (Polly, 1999)

The first Sydney Mardi Gras was organised in 1978 in solidarity with gays and lesbians in San Francisco who were fighting against a homophobic bill proposed by Republican Senator John Briggs. Activists in a range of community and political organisations, including the Active Defence of Homosexuals on Campus, the Gay Task Force and Campaign Against Moral Persecution (CAMP), formed a committee to organise a rally. On 24 June 1978 a daytime rally was held, followed by a night-time carnival. According to Ken Davis (quoted in Begg, 1999):

> City shoppers and workers saw an unprecedentedly large lesbian-led street march. The march passed without incident, everyone exhilarated by the turnout and the vehemence of our demands against discrimination, the law and violence. For many, it was their first demonstration, their first coming out.

However, the night-time carnival was marred by conflict with police who prevented a march by demonstrators into Kings Cross, Sydney's red light district with fifty-three people being arrested.

In response to the police actions a protest rally was held the following 15 July, with more than 2,000 people marching to demand that the NSW Labour government drop the charges. At the time it was the largest lesbian and gay rally ever held in Australia. After marching through Kings Cross, the demonstrators stopped in front of Darlinghurst Police Station and laid wreaths of pansies. In response, the police arrested eleven more people. The police actions provoked national outrage with rallies in support of the marchers also being organised in Melbourne, Adelaide and Brisbane. A commemorative rally held the following year attracted 3,000 demonstrators. Mardi Gras had become an annual celebration of gay and lesbian pride which marks broader changes in Australian and New South Wales society. As Bucknell records:

> There are Mardi Gras regulars who weren't born when the first parade, a demonstration for gay and lesbian rights, was held in June 1978. 'Out of the bars and into the streets,' the protesters chanted along Darlinghurst's Oxford Street. Many were bashed and arrested as they tried to escape a police blockade. How far we've come. For a decent spot to view the parade along the Golden Mile these days you need to set up several hours early with a milk crate (street price: A$25) to beat the 500,000 others. The police work with Mardi Gras marshals along the route.
>
> (1999)

However, from the 1980s on there have been sharp debates within Sydney's gay and lesbian community over the direction of the Mardi Gras and the extent to which it should reflect political actions against its more commercial, party emphasis. For example, lesbians, although always present from the first Mardi Gras, were secondary players up until the late 1980s. Fierce internal debate accompanied the proposal to incorporate the word 'lesbian' into Mardi Gras' name in 1988. This commitment to 'coalition politics' reflected a growing interest from lesbians in the celebration of sexuality that Mardi Gras had become during the 1980s (Bucknell, 1999). The AIDS epidemic has also added another dimension to the Mardi Gras with the significance being recognised not only of a safe sex message but also remembering those who have died from AIDS. According to Begg (1999), those who wanted to make Mardi Gras more of a party and less of a demonstration gained the upper hand. In 1982 Mardi Gras was moved to a summer schedule and Brian McGahen, on behalf of the organising committee, declared 'we are keen on having maximum commercial participation'. To justify the shift in emphasis in Mardi Gras, the 1983 organising committee explained, 'We do not see politics in any narrow sense. Our right to lead our chosen lifestyle is a major political demand' (in Begg, 1999). Raising lesbian and gay visibility has become a focus of Mardi Gras, rather than just using the march to campaign for equal rights. Nevertheless, as activist Craig Johnston observed, the transformation from political movement to mainstream consumer event may have been inevitable. 'Same politics of dignity and rights, different style. Gay liberation's ideological sharpness was defused as it diffused into the scene queen's body' (Bucknell, 1999).

Today, more than 500,000 people participate in the Mardi Gras parade whether as participants or spectators. In addition, many people watch the parade on television. In 1985 the media were inflammatory. The *Sydney Morning Herald*, which now produces its own Mardi Gras lift-out, was far from supportive. In the 1985 *Herald* report on the parade a journalist invented quotes she attributed to the then Mardi Gras director. To his horror he read, on the Monday after the parade, references he was supposed to have made to people with AIDS as 'freaks' and 'Elephant Men' (Wherrett, 1999). The *Herald* was forced to publish an apology (Bucknell, 1999).

Research by the Australian Graduate School of Management (AGSM) in 1998 found that Mardi Gras contributed more than $40 million to the Sydney economy. There were more than 5,000 international visitors to Sydney during the 1998 festivities, 3,600 of whom came specifically for the event. Of the 7,300 interstate tourists, 4,800 had Mardi Gras in mind, and AGSM also recorded 2,400 'holidays at home' (Pettafor, 1999). According to Begg (1999), Qantas earned $1.5 million from international visitors, and the Mardi Gras attracts more international and interstate visitors than any other cultural festival in Sydney, Melbourne, Perth or Adelaide with the festival guide carrying advertising from Hahn Ice, Qantas, Telstra, Foxtel, Land Rover and other large corporations.

The pink dollar is now a significant attraction to companies. In 1999 the 25th Sydney Gay and Lesbian Mardi Gras attracted a record A$800,000 worth of corporate sponsorship. Telstra, Qantas, Coca-Cola, Lion Nathan and Southcorp Wines were among the blue chip business names that have lined up to support and cash in on the annual celebration of gay and lesbian sexuality. For example, Telstra launched a commemorative Mardi Gras phone card featuring its official Drag Queen spokeswoman, Ms Candee. Coca-Cola used the festival to promote its Mount Franklin Bottled Water with full page advertisements in the gay press featuring a Sydney Drag Queen, Verushka Darling. Stolichnaya Vodka sold its Lemon Ruski bottles across Sydney with tiny pink feather boas wrapped around their necks (Hornery, 1999). However, the degree of commercialism and corporate involvement attached to Mardi Gras has served to heighten tensions surrounding the meaning of the event within the gay community. For example, Ian Johnson, principal of Significant Others, a marketing consultantcy specialising in targeting gays and lesbians stated:

> I don't think that corporate sponsorship is necessarily a bad thing. However, I think that by it's very nature whatever deals are struck have to be handled very sensitively and strategically ... At the end of the day the interests of the gay and lesbian community must be paramount otherwise they [the sponsors] risk the support of Mardi Gras's own membership base.
> (quoted in Hornery, 1999)

More vociferous were the comments of Anthony Yeo, in response to the prevention of a gay group collecting donations at a Mardi Gras Fair Day, in a mailing to several gay and lesbian news lists:

> What on earth has gotten into the very swelled heads that permeate Mardi Gras? First The Rainbow Party, now this insanity. Rebel people rebel, against a commercially driven organisation that has completely lost all touch with reality and the community it is supposed to serve. It comes as no surprise no-one from the Erskineville bunker will comment, far too afraid of the backlash?

The same message also contained a previous mailing from Norrie in relation to the same issue:

> That's just MAD! Does anyone think these people haven't gone too far? Stop giving them your money! Stop supporting their fascism with your cash! Don't buy a ticket to the Mardi Gras party! Next year, if you've already forked out!
>
> Starve the bastards! The community festival will live on, and the parade, with or without the sponsorship dollars! This gig started as a community driven event, and it may need to again for the sake of the community!

Although corporate sponsorship is now an important part of the Mardi Gras, the extent to which such commitment extends is perhaps debatable. For example, the front page Column 8 of the *Sydney Morning Herald* (1999) reported one reader's comments:

> AT LAST ... Qantas has acknowledged that the Sydney Gay and Lesbian Mardi Gras is a major event by mentioning it in its in-flight magazine. But

> although Qantas is a sponsor and official carrier, there is no mention of it being a gay and lesbian event. It's described thus: OOH LA LA, IT'S MARDI GRAS – Sydney's wildest, brightest and most fabulous festival turns 21 this year ... Some tourists might get a bit of a surprise.

The tensions surrounding the meaning of Mardi Gras, however, are well recognised by the festival's organisers. At the official opening of the 1999 Gay and Lesbian Mardi Gras on the Sydney Opera House forecourt in front of an audience of more than 20,000 people, Festival President David McLachlan used a wide-ranging speech to acknowledge the Aboriginal traditional owners of Sydney, applaud the character of French tennis player Emily Mauresmo, criticise the conservative nature of all politicians on social reform, and criticise the Opera House Trust, which had withdrawn permission for the Sisters of Perpetual Indulgence to conduct tours of the site during the festival. According to Mclaughlin, the festival marked a special time. '... when we celebrate and illuminate and share our lives, not only with each other, but with the whole bloody world' (ABC News, 1999). The chief executive officer of the AIDS Council of New South Wales, Robert Griew, said at the opening that the Mardi Gras is also important as a reminder of how much remains to be done:

> It started as a political struggle, and I guess for us it means we think about all the people we've lost with HIV, and the fact that the struggle keeps going on ... It's not over, it continues, and it's a time to remember that, but to celebrate the strength of the community, and the base that is for the struggle with HIV.
>
> (ABC News, 1999)

David McLachlan's reflections on Mardi Gras in an article with a sub-heading, 'Fairy stories do come true', provide an opportunity to witness his own observations of the meaning of Mardi Gras in a manner that reflected the various meanings associated with the annual festival. As with many Australian gays and lesbians the Mardi Gras was, and continues to be, an important defining moment in life, providing an opportunity to come out publicly and be able to celebrate their sexuality in an accepting environment (see also Wherret, 1999). The ritual nature of carnivals and festivals, such as the Mardi Gras, as liminal space is highlighted in McLachlan's (1999) observation: 'For many it is the ritual of Mardi Gras which is its most defining quality. It is the annual cycle of anticipation, mounting excitement, repeated we-do-it-every-year events, the big day itself and then the letdown which is an important part of the ebb and flow of their lives.' McLachan's comment on the Sydney Mardi Gras provides an opportunity for meaningful inversion, in the same way as carnival does in Latin and South America, the nature of that meaning is complex and multi-layed even seen from within the gay and lesbian community.

> For many it is Mardi Gras as an arts, cultural and community festival which is most valued and most important. The festival is now one of the major arts festivals in Australia with an international reputation. It is an opportunity to experience the best of artistic expression as part of a curated lesbian and gay arts festival. For others it is Mardi Gras as a community cultural arts festival that is most important. It is an opportunity for our various and often quite disparate communities to come together to share, inform, entertain and engage with each other and with a wider community with artists and performers from those communities providing the lead and the focus. Many will say to you that it is the public scale and dimension of the festival and in particular the parade that is most important. The notion of taking to the streets is seen as empowering. The launch occurs in the very centre of one of Sydney's prime civic spaces. Our parade for the best part of a day overwhelms central Sydney. For much of the final week most of inner and central Sydney (and increasingly further afield) is a buzz with queer tribes from everywhere publicly and loudly celebrating the spirit of Mardi Gras. For others the absolute core of Mardi Gras is its political purpose. It grew from a time when taking to the streets was an intensely political act. It has allowed us more effectively than any other means to deliver a message about homophobia, intolerance and legal equality. Above all it is about visibility which in 1999 is still capable of informing, affirming and challenging.
>
> Many say that Mardi Gras is increasingly about money. More and more it is seen as having economic importance for the Sydney and Australian economy as a whole rather than just for Oxford Street businesses. At the same time it remains the most important fund-raising opportunity for many of our community organisations. The economic benefits that flow to businesses within and outside our community are likely to continue to grow. Some see this as Mardi Gras' greatest challenge: to remain true to its community and political purpose in the face of its increasing attraction to commercial interests. Others see this as the ultimate path to real change and real acceptance in a city where money talks and opens doors and can ultimately deliver real social and political change.
>
> (McLachlan, 1999)

However, McLachlan (1999) concludes by noting that for him Mardi Gras is primarily a deep personal significance:

> It still stimulates in me that childlike excitement which has me counting down the days and leaves my stomach fluttering with anticipation. Often times it exceeds my highest expectations. Sometimes it must be said it disappoints me. Invariably it leaves me drained and exhausted and sad that it is over.

The hosting of 21st Mardi Gras in 1999 provided an opportunity for much reflection on the meaning of Mardi Gras. A series of responses to the question of meaning is extremely revealing. To political satirist Pauline Pantsdown, Mardi Gras is 'A big chance to broaden the definition of mainstream Australia and of the gay and lesbian scene itself.' According to Sister Salome of the 9th Mystic Rhinestone of the Order of Perpetual Indulgence, 'For the sisters, it means hard work, high visibility and, more often than not, wimple rash.' For Peter Baldwin President of Gay and Married Men's Association and a father of three, Mardi Gras has a wider meaning, 'Reflection on the fact that society still practises discrimination, not just against gays and lesbians but a lot of minority groups', while for Janet Carter, workshop artist for Mardi Gras, it represents, 'A chance for me to have my politics with a bit of icing on top. You can make big statements about homosexual issues such as equal rights and HIV in a really fabulous way.' This integration of politics and party is also seen in the comments of writer and historian Garry Witherspoon, 'Hedonism and optimism. It means summer in Sydney. But it also still means politics ... we need equality in same-sex relationships, equal age of consent' (all quoted in McDonald-Leigh, 1999).

The Sydney Mardi Gras, described as a 'sequinned revolution' (Wherrett, 1999), has undoubtedly provided a space within which many different levels of meaning occur. It is also a space of belonging and identity as well in which some gays may feel more at home than in their normal home environments (Hughes, 1997). Nevertheless, representation and control of this space remain contested, not only by conservative elements in society, but also within the gay and lesbian community. Begg (1999) posed the question as to whether visibility was enough. Visibility is undoubtedly important. Mardi Gras is a powerful 'in your face' reminder that there are hundreds of thousands of people who refuse to be silenced by homophobic attitudes. As the 1999 program stated, 'We won't stand for any infringement of our basic right to be who we want to be and love who we want to love' (Begg, 1999). For Begg along with others, visibility is not enough:

> Although it may be 'OK to be gay' for the one night of Mardi Gras, when the party is over discrimination against gays and lesbians remains. As the 1999 festival guide points out, gay sex is still illegal in 40 per cent of countries around the world ... In Australia, there are a myriad of laws which discriminate against gays and lesbians. Today, Mardi Gras is part of establishment culture. It is an 'outrageous' night which shocks the shockable (Fred Nile maintains his vigil against the sin of it all), but simultaneously makes private businesses millions of dollars and provides a platform for hypocrites such as Kim Beazley, Bob Carr and Peter Collins. It is one big queer night during which everyone is expected to wear as much (or little) leather, sequins and latex as possible, so long as they come down after the party and go home. To achieve real equality for lesbians and gays, much more is required. The radical movement which sparked the first Mardi Gras provides the clues as to what is required.

Begg's comments will strike a chord for many who are seeking equal rights for gays and lesbians. However, they represent just one strand of several in attaching meaning and identity to the Mardi Gras. The Mardi Gras, as with other forms of gay travel, is significant for the relationship between sexuality and tourism in its broadest sense. It demonstrates the wide variety of meaning and association attached to the event by a number of members of the community, it also indicates the difficulty that some members of mainstream society may have in accepting such expressions of sexuality and, therefore, identity, in relation to travel. The complexities of understanding the relationship between sex and tourism in the context of gay travel are therefore no different from heterosexual expressions of this relationship, except that, perhaps, from the viewpoint of some in Western society, the gay and lesbian community lies at the margins of the margin. Nevertheless, the tensions created in the liminal space by the Mardi Gras and other gay events and travel products may have created opportunities for better understanding of gay and lesbian issues. Visibility may therefore be a precursor to understanding and eventually equality. Hopefully, the myths associated with gay tourism and its relationship to paedophilia and child sex tourism may eventually be dispelled. The marginal space of gay tourism is therefore very similar in structure to the marginal spaces of heterosexual tourism. In both cases issues of identity and meaning occur. However, it is possible that because of homophobic elements in Western society tourism may be more important for identity for gays than for heterosexuals (Hughes, 1997). If this is the case, then Mardi Gras is as much a cause for celebration of tolerance and understanding in a multi-cultural, multi-lifestyle and diverse society than it is protest. To conclude this chapter with a life-affirming comment from Bucknell's (1999) review of Wherrett's (1999) book on twenty-one years of Mardi Gras:

> One of my favourite stories is from Greg Logan who, in 1996, walked alone in the parade carrying a placard saying 'Mum ... there's something I've got to tell you'. He warned his parents to watch the telecast, choosing this unconventional way to break their silence, and it worked. That's Mardi Gras: non-conformist, ever-changing and always inyerface.

6 Sexual slavery

Trafficking sex

Trafficking in persons – the illegal and highly profitable transport and sale of human beings for the purpose of exploiting their labor – is a slavery-like practice that must be eliminated ...

Trafficking in persons is a profound human rights abuse, and women are particularly vulnerable to this practice due to the persistent inequalities they face in status and opportunity.

(Ralph, 2000. Opening and closing remarks by Regan E. Ralph, Executive Director, Women's Rights Division, Human Rights Watch, International Trafficking of Women and Children, in her testimony before the United States Senate Committee on Foreign Relations Subcommittee on Near Eastern and South Asian Affairs, 22 February 2000.)

Historically, as demonstrated in Chapter 1, Western tourism has been a highly gendered activity. Until the development of industrialised tourism in the mid-nineteenth century – women generally stayed at home, men travelled. Mass tourism is generally acknowledged to have commenced on the 5 July 1841, when the first conducted excursion train of Thomas Cook left Leicester station in northern Britain. This seemingly inauspicious event in fact heralded the beginning of the era of cheap travel. Cook's tours were attractive to the rapidly growing middle classes of newly industrialised Britain, offering the romance of travel from a position of safety. Cook provided insulation from beggars; he speeded up border crossings; he booked accommodation and transportation in advance; and he eliminated the hassles of foreign exchange. Moreover, Cook allowed women the opportunity to travel and his services were especially popular with Victorian women, especially single women (Swinglehurst, 1982).

The reasons for the gendered nature of travel lie within the gendered nature of Victorian England, a woman's place was more than in the home, women were also regarded as being the property of their husbands or fathers. The provision of chaperoned, guided tours by Thomas Cook therefore provided a space in which male control was shifted from the home to the train carriage but could still be maintained. Control was also maintained through culturally sanctioned myths in which women travellers could be in danger of losing their chastity, or, even worse be lost to the white slave trade of the Arab world

(Blanch, 1984). Such myths stay with us to the present day. Yet, behind the myths are harsh realities, women do go missing or are raped when travelling, indeed stories of female hitchhikers being kidnapped and sexually assaulted have become one of the recurring images of the late twentieth-century crime writing genre. Female travellers are therefore often caught in the space between the desire for the freedom of travel and the threat of sexual violence – either real or imagined – both of which act to constrain if not define the social space of travel. Whether a female is more likely to be attacked while travelling than in her own home is a moot point. In some societies, such as that of the authors, women are more likely to be sexually assaulted by someone they know. This does not make travel spaces any less threatening, however. There are numerous suggestions in the literature that the commoditisation of the body in tourism promotion and advertising may lead to an increase in attacks on females, whether locals or tourists. For example, Jones' (1986) study of the development of the Gold Coast, Australia's leading mass tourism destination, characterised the resort as 'A Sunny Place for Shady People'. The high rates of crime in the region, which was described as 'the State's rape capital' (Kyburz in Jones, 1986: 110), were regarded by Jones as being closely related to the inherent nature of the local tourism industry which utilised bikini clad women and the promise of the availability of such women extensively in the destination's promotion. Similarly, Bishop and Robinson (1998) in their study of sex tourism in Thailand point to the role of advertising and the representations of certain images of Thai women as a contributing factor in male tourist attitudes towards Thailand and Thai women. Although such representations are extremely important in attempting to understand the multi-layered dimensions of sex tourism, this chapter focuses on the role to which sex slavery and trafficking are related to the development of sex tourism. As noted in Chapter 2, by becoming a sex tourist, by responding to the imagery of sexual service, that tourist might be said to be transgressing from the licit to the illicit. The illicit may go so far as to include the forced provision of sexual service to the tourist. In this situation not only does the tourist clearly now begin to share the possibility of condemnation, but so also do the institutions and individuals which allow such a situation to exist. The remainder of this chapter will provide an overview of the issues that pertain to sex slavery and trafficking and additionally make reference to paedophilia. In terms of the paradigms described earlier this relates to overt exploitation for profit and the denial of human dignity to those who are the victims. It represents an extreme in the model advanced of sex tourism, but yet also represents links to wider societal norms. At first sight it can be argued it represents the extreme whereby women and children are simply commodities to be purchased, and like other goods have no rights. However, it will be contended that it is not simply a matter of age and gender that is at issue, but essentially a question of economic as well as cultural power. The sex slavery and trafficking discussed here are part of a wider exploitation of the poor. In terms of the conceptualisation of marginality previously used – the victims here might be regarded as those at the very margins of social and political processes condemned by those processes to a poverty which leaves them defenceless against unprincipled exploitation.

Trafficking and slavery

Under the US International Trafficking of Women and Children Victim Protection Act of 1999 – 'trafficking' is defined as:

> the use of deception, coercion, debt bondage, the threat of force, or the abuse of authority to recruit, transport within or across borders, purchase, sell, transfer, receive, or harbor a person for the purpose of placing or holding such person, whether for pay or not, in involuntary servitude, slavery, or slavery-like conditions or in forced, bonded, or coerced labor.
>
> (US Congress, 1999)

According to Human Rights Watch (1999), a US based non-government organisation, 'Trafficking, the illegal and highly profitable transport and sale of human beings across or within international borders for the purpose of exploiting their labor, is a human rights abuse with global dimensions' with many thousands of women and girls around the world being lured, abducted or sold into forced prostitution, forced labour, domestic service, or involuntary marriage. Trafficking is therefore closely related to the wider issue of sexual slavery (Rosenfried, 1997; Matsui, 1999). Although sympathetic, this chapter does not take the view of Barry (1984: 40) that:

> Female sexual slavery is present in ALL situations where women or girls cannot change the immediate conditions of their existence; where regardless of how they got into those conditions they cannot get out; and where they are subject to sexual violence and exploitation.

Nevertheless, Barry does point to the significance of the exploitation that the loss of individual control brings in many situations where prostitution exists.

Many well-publicised media accounts of sex tourism in Thailand in particular have noted the extent to which women and girls have been bonded into prostitution often through agents and brothel owners making loans or payments to relatives (e.g. see Bishop and Robinson, 1998; Matsui, 1999). However, the selling of women into prostitution is not isolated to Thailand – it is a global phenomenon in which the female body is objectified into a commodity to be bought and sold – it is the (il)logical extent of the objectification of labour noted in Chapters 1 to 4 in which not only are humans seen purely as a unit of sexual labour which is under the control of their 'owner', but it also represents the denial of self of one human by another.

The Executive Director, Women's Rights Division of Human Rights Watch argues that the number of persons trafficked each year is impossible to determine, but it is clearly a large-scale problem, with estimates ranging from hundreds of thousands to millions of victims worldwide (Ralph, 2000). The International Organization for Migration has reported on cases of trafficking in South-east Asia, East Asia, South Asia, the Middle East, Western Europe, Eastern Europe, South America, Central America, and North America. For example, the US State Department estimates that each year, 50,000–100,000 women and children are trafficked into the United States. Many women are trafficked to work in brothels; about half are trafficked into bonded sweatshop labour or domestic servitude. Once in the United States, the women who work in brothels typically are rotated from city to city to evade law enforcement, keep the women disoriented and give clients fresh faces (Rosenfried, 1997). In her testimony before the Senate Committee on Foreign Relations, Ralph (2000) reported that:

> In August 1999, a trafficking ring was broken up in Atlanta, Georgia that authorities believe was responsible for transporting up to 1000 women from several Asian countries into the United States and forcing them to work in brothels across the country.

Trafficking patterns

Human Rights Watch's documentation of trafficking in women found that while the problem varies according to the context, certain consistent patterns emerge. In all cases, the coercive tactics of traffickers, including deception, fraud, intimidation, isolation, rape, drugs, threat and use of physical force, and debt bondage, are at the immediate core of the trafficking problem. According to Ralph (2000):

> In a typical case, a woman is recruited with promises of a good job in another country or province, and lacking better options at home, she agrees to migrate. There are also cases in which women are lured with false marriage offers or vacation invitations, in which children are bartered by their parents for a cash advance and/or promises of future earnings, or in which victims are abducted outright. Next an agent makes arrangements for the woman's travel and job placement, obtaining the necessary travel documentation, contacting employers or job brokers, and hiring an escort to accompany the woman on her trip. Once the arrangements have been made, the woman is escorted to her destination and delivered to an employer or to another intermediary who brokers her employment. The woman has no control over the nature or place of work, or the terms or conditions of her employment. Many women learn they have been deceived about the nature of the work they will do, most have been lied to about the financial arrangements and conditions of their employment, and all find themselves in coercive and abusive situations from which escape is both difficult and dangerous.

Debt bondage is regarded by Human Rights Watch (HRW) as the most common form of coercion (also see Matsui, 1999). In this situation women are told that they must work without wages until they have repaid the purchase price advanced by their employers, usually this amount far exceeds the cost of their travel expenses. In addition, the amount of debt is routinely augmented with arbitrary fines and/or dishonest account keeping. Employers also often maintain their power to 'resell' indebted women into renewed levels of debt. Sometimes the size of debt is such that it can never be fully repaid. To prevent escape, employers also take advantage of the women's vulnerable position as international or domestic migrants. In order to further coerce women, their employers may also make threats against the woman's family. Often they do not speak the local language, are unfamiliar with their surroundings, know no-one apart from their employer or their fellow workers, and fear arrest by local law enforcement agencies. In many countries government attempts to combat trafficking in persons have been entirely inadequate and, where policies exist, agencies' efforts are often ineffective (Matsui, 1999). Kristof (1996) reports the case of a 15-year-old Vietnamese girl who was sold to a Vietnamese brothel by a man who effectively kidnapped her. The girl's mother finally tracked her daughter down but since the brothel owner had paid for the girl – even if to a kidnapper – the mother could not take the girl home. Instead, the mother had to settle for signing a contract with the brothel owner, stipulating that when the girl earned back enough money she would be returned to her family. According to an anonymous Cambodian journalist, who has written about the problems of child prostitution and who is cited by Kristof (1996):

> If the mother tried to grab her daughter and take her out of the brothel, the owner would have them beaten up … And if the mother takes on the brothel owner, she can't win. The brothel owner can just pay some money to the police, or give the girl to the police, and the parents will lose.

This situation reflects the testimony of Ralph (2000):

> In many cases, corrupt officials in the countries of origin and destination actively facilitate trafficking abuses by providing false documents to trafficking agents, turning a blind eye to immigration violations, and accepting bribes from trafficked women's employers to ignore abuses. We have even documented numerous cases in which police patronized brothels where trafficked women worked, despite their awareness of the coercive conditions of employment. And in every case we have documented, officials' indifference to the human rights violations involved in trafficking has allowed this practice to persist with impunity. Trafficked women may be freed from their employers in police raids, but they are given no access to services or redress and instead face further mistreatment at the hands of authorities. Even when confronted with clear evidence of trafficking and forced labor, officials focus on violations of their immigration regulations

> and anti-prostitution laws, rather than on violations of the trafficking victims' human rights. Thus the women are targeted as undocumented migrants and/or prostitutes, and the traffickers either escape entirely, or else face minor penalties for their involvement in illegal migration or businesses of prostitution.

Several case studies of trafficking in women can be offered. From 1994 to 1999, Human Rights Watch carried out an extensive investigation of the trafficking of women from Thailand into Japan's sex industry, interviewing numerous trafficking victims directly. The women had been recruited for work in Japan by friends, relatives, or other acquaintances, who told them about high-paying overseas employment opportunities. Upon their arrival in Japan, the women were delivered to brokers who sold them into debt bondage in the sex industry. While in debt, they could not refuse any customers or customers' requests without their employers' permission. HRW also interviewed a woman who was promised a job in a Thai restaurant in Japan, but instead was taken to a bar where the other Thai 'hostesses' told her she would have to work as a prostitute. She recalled, 'They told me there was no way out and I would just have to accept my fate. I knew then what had happened to me. That first night I had to take several men, and after that I had to have at least one client every night' (in Ralph, 2000; Matsui, 1999 offers several similar examples).

According to Human Rights Watch, thousands of Thai women are trafficked into forced labour in Japan each year, often with the involvement of the Japanese-organised crime syndicates (*yakuza*) and with the Japanese and Thai governments failing to respond even though they are aware of the problem. Matsui (1999) regards Japan as the most common destination for women who have been sold into sexual slavery. In her analysis of the trafficking of women in Asia, Matsui notes that one of the reasons for the extent of sexual slavery in Japan is the size of its sex industry which generates over four trillion yen (US$30 billion) per year, 'a sum equal to the total of Japan's national defence budget, or 1 per cent of the gross national product (GNP). The sex industry compensates for the shortage of young Japanese women by importing Asian women, who are even cheaper to hire' (Matsui, 1999: 19).

There are areas in towns in the Nagano and Ibaragi Prefectures where hundreds of Thai women live – leading to the nicknames of 'Little Bangkoks' – although significant numbers of Filipino women have also been trafficked to Japan as well. According to Matsui (1999: 18), the women represent important financial assets for the yakuza: 'a *yakuza* group that has ten Thai women in its control can gain a billion yen [US$7 million] per year in profits'. Several reasons can be provided for the expansion of Japan's sex industry. First, the continuing influence in Japanese society of a sexual climate derived from the licensed prostitution system of the feudal era (Matsui, 1999) with Japan's first anti-prostitution law not being enacted until 1956 (Robins-Mowry, 1983). Second, a corporate culture, in which women are seen as a source of 'comfort' for the 'company warrior' (Matsui, 1999: 19). Indeed, such was the significance of prostitution in Japan's male-dominated culture that, following both the 1911 and 1923 earthquakes, the reconstruction of licensed prostitution quarters took priority over the rebuilding of schools (Robins-Mowry 1983: 248). Third, there was a dramatic increase in the number of women sent to Japan in the early 1990s following increased opposition to Japanese sex tours in Thailand and the Philippines. In short, the Thai sex industry business, faced with a potential loss of demand responded by 'exporting' women to meet demand in the demand-generating country. This has not been restricted to Japan, and examples of similar practices have also been observed in countries like Australia and New Zealand.

Japanese traffickers and employers have traditionally faced little fear of punishment. If arrested at all, they have been charged by the Japanese authorities only with minor offences for violations of immigration, prostitution, or entertainment business regulations. It is estimated that 3,000 Thai women each year manage to escape from their conditions and make their way to the Thai Embassy in Tokyo (Matsui, 1999). However, the trafficked women remain the focus of the authorities' actions with the women being detained in immigration facilities following arrest and then are summarily deported with a five-year ban on re-entering Japan. According to Ralph (2000) 'Trafficking victims have no opportunity to seek compensation or redress, and no resources are provided to ensure their access to medical care and other critical services.' In short, rather than being treated as victims they are seen as part of the problem.

The Thai government's actions are also ineffective even though they have adopted laws in order to combat trafficking. For example, the passport applications of women and girls ages 14 to 36 are subject to special scrutiny, and if investigators suspect that a woman may be going abroad for commercial sexual purposes, the application is rejected. As Ralph (2000) observes, this policy, even though well intended:

> trades one human rights problem for another by discriminating against women seeking to travel and limiting their freedom of movement. It also makes women who want to migrate even more dependent on the services of trafficking agents, because it is difficult for women to obtain travel documents by themselves.

In addition to the trafficking of women from Thailand to Japan, Burmese (Myanmar) women are trafficked to Thailand to work in brothels, often with the complicity of Thai officials. According to Matsui (1999: 21).

> Raids on brothels and the extension of compulsory education to include middle school have resulted in a slight decrease in the number of Thai girls sold into prostitution. However, to compensate for the decreased availability of Thai girls, [traffickers] have come to target young girls in neighbouring countries.

It is estimated that over 40,000 girls from Myanmar have been imported into Thailand with many ending up at the tourist resorts of Chiang Mai and Pattaya, although 'the largest number of enslaved Burmese girls are concentrated in Ranon, a fishing port near the southern Thai border, where they are said to be confined in more than forty brothels surrounded by electric fences' (Matsui, 1999: 21). Similarly, girls from minority tribal groups in Yungnan Province in Southern China are also being imported into Thailand with many then being sent on to other locations.

Vietnam

Vietnamese authorities are also struggling to control the booming sex trade in the country's urban centres (Huckshorn, 1996; Watkin, 1999, 2000). Of Vietnam's registered commercial sex workers, 15 per cent are aged under 18, according to the Ministry of Labour, Invalids and Social Affairs, with the highest rate of teenage prostitution occurring in southern Vietnam. However, Nguyen Thi Hue, who heads campaigns to eradicate social ills in Vietnam, was quoted in 1999 as saying that the Vietnamese police had files on some 185,000 prostitutes nationwide, and that in 1998 some 30 per cent of all sex workers in this country of 79 million people were thought to be under the age of 16 (Reuters, 1999; Solomon, 1999). Since 1997, teenage prostitutes have not faced criminal charges but instead have been sent to rehabilitation centres. Nonetheless, according to a Ministry of Labour report, nearly 80 per cent of those apprehended in 1998 have returned to prostitution (Watkin, 1999). Furthermore, an estimated 20 per cent of Vietnam's commercial sex workers are held in brothels against their will. Urban incomes average four times those in rural areas, which has created an enormous economic underclass and drives thousands of young women from the countryside to the big cities. However, pay for the unskilled is so low they struggle to make ends meet, let alone provide for their often desperately poor families. According to Watkin (2000), authorities are often quick to blame foreigners for creating the demand for commercial sex, although a recent government report showed that 60 to 70 per cent of clients are state employees. Official complicity, together with a pervasive denial of the problem, serves to reinforce the exploitation of the sex workers. According to a representative of one organisation involved in assisting the sex workers:

> Prostitution is condemned by the whole of society and prostitutes are terrified that their families will discover and abandon them because of what they do for a living. Many want to stop, but the bosses threaten to tell their families and so they just keep sending money home – it's a never-ending circle. Their friends and families don't know, or don't want to know, what they actually do in the cities. All they know is that the money is good.

As well as its own domestic sex industry, Vietnam is also increasingly becoming a major source for the trafficking of young girls into neighbouring Cambodia and China. In Vietnam, women can be sold to brothels for two million dong (HK$1,120), but according to the Vietnamese Ministry of Labour, brothels in Taiwan and China pay up to US$7,000 (HK$54,250) for young Vietnamese women (Watkin, 1999). In addition, the regional economic downturn in Asia in the late 1990s was further fuelling the trafficking of women and children, which has flourished as a result of poor law enforcement and corruption.

The growth of sex tourism-related child prostitution has attracted substantial attention from international aid agencies and non-government organisations. Christine Beddoe, tourism programme director for international group End Child Prostitution, Pornography and Trafficking (ECPAT) Australia, said there was credible evidence to suggest foreign child sex tourists were active in Vietnam: 'There is really strong anecdotal evidence coming from Hoi An where it appears foreign females are abusing underage boys' (in Solomon, 1999). Beddoe added that the northern mountain resort of Sapa was also being repeatedly mentioned from different sources who remark on child sex abuse, mostly with young girls from the Hmong ethnic minority. However, Beddoe stated, 'I have absolutely no doubt the Vietnam Administration of Tourism supports a wider national campaign on the prevention of sex tourism including child sex tourism' (in Solomon, 1999).

There are no reliable statistics on child prostitution in Vietnam, but the police reports indicate that the number of sex crimes in general marked annual increases in recent years. At the end of 1999 some nineteen countries operate extraterritorial laws that can be used to convict people of child sex crimes committed on foreign soil. Up to the time of writing (April 2000) neither Hanoi nor foreign courts have prosecuted foreign nationals for child sex crimes that occurred in Vietnam. In addition, attempts by Vietnamese authorities to raise the profile of the role of sex tourism in Vietnam have been opposed by many in the tourism industry. A Vietnam News Agency (VNA) report on tourism released in September 1997 highlighted some of the negative implications of tourist growth. According to the report 'many foreign tourists were coming to Vietnam to satisfy their sexual desires or engage in women-trafficking' (in A. Edwards, 1997). However, Edwards noted that tourist industry executives reacted angrily to the report: 'What do they think this is here? A James Bond movie?' a hotel executive in Ho Chi Minh City said by phone. 'I don't think officials understand the ramifications of stories like this … These articles, when they reach foreign shores, have a very adverse effect on tourism.'

Bosnia

The trafficking of women and girls is not just confined to Asia. Since the end of the war in Bosnia and Herzegovina in 1999, thousands of women have also been trafficked from Eastern Europe and the former Soviet Union to Bosnia for forced bonded prostitution. This occurred despite the presence of Joint Commission Observers and the International Police Task Force. It is perhaps ironic that although so much attention was focused on the use of rape as a weapon of war and ethnic cleansing during the break-up of Yugoslavia, little attention has been focused in the Western media on the trafficking of women and girls. Human Rights Watch (1999) even found evidence 'that some officials were actively complicit in these abuses, participating in the trafficking and forced employment of the women and/or patronising the brothels'. The situation was compounded by what occurred when the authorities did take action. As in the case of Japan, the women were the ones treated as criminals. They would be arrested, fined for their illegal status as immigrants and prostitutes and then deported.

However, in early 1999, 'deportation' in the Bosnian context – a country without an immigration law – translated into being dumped across a border. From the Federation, women found themselves dumped in Republika Srpska. And vice versa. This pseudo-deportation scheme only facilitated the trafficking cycle. Women 'deported' across the internal borders without money, family contacts or other resources could be quickly picked up and re-sold (Ralph, 2000).

The pattern of trafficking reflected in the Bosnian situation repeats that of Asia and many other parts of the world, that is, it tends to be from poor areas to richer regions, an issue which will be returned to in the next chapter, while an additional dimension to trafficking is the increasing extent of trafficking in children.

Trafficking children

According to the International Labour Organisation (ILO) (1995) there are estimates of at least 1 million Asian children forced into different forms of 'sexual exploitation' with the problem being especially 'alarming' in seven Asian countries: Korea, Thailand, the Philippines, Sri Lanka, Vietnam, Cambodia and Nepal:

> They can rarely seek help or be reached, because child prostitution is criminal by nature and hidden from public view. Case studies and testimonies of many children in prostitution speak of a trauma so deep that the child is unable to return or re-enter a normal way of life.

Similarly, Kristof (1996) noted, that more than a million girls and boys, aged 17 and younger, are engaged in prostitution in Asia:

> [A]lthough all these figures are no more than wild guesses. The slaves, in the sense of those who are locked up or owned by a brothel, are a minority of the total. But even among those teen-agers who now offer themselves on

> street corners, many first entered the sex trade unwillingly, sold by parents or simply kidnapped off the street.

The ILO has been making a firm commitment to ending the involvement of children in prostitution for many years with this commitment being reflected in Convention 29 on forced labour adopted in 1930 and other ILO standards. More recently, the ILO has developed an International Program on the Elimination of Child Labor (IPEC) which has been operating in Thailand and the Philippines. In Thailand forced child prostitution has involved 'regional trafficking' of children from the poorer areas of the northern hill-tribes and, more recently, the problem has extended into trafficking across the Thai borders from Burma and Cambodia. According to an ILO worker, 'Ongoing action includes legal action against the traffickers and brothel owners and procurers, the provision of educational opportunities and alternative education, skills training, income generation and awareness-raising among the children, parents and community leaders on the effect of forced prostitution.'

According to the Movement to Prevent Child Prostitution (MPCP) (1996) Sri Lanka, along with Thailand, the Philippines, Taiwan and Vietnam, has become notorious as a destination for paedophiles. The estimated number of young Sri Lankan boys (aged between 6–14) who have been victimised by foreign paedophiles is in the region of 10,000–15,000 (United Nations Children's Fund – UNICEF) with some reports claiming up to 30,000 (End Child Prostitution in Asian Tourism – ECPAT) in the early 1990s (cited in MPCP, 1996). More recently, there has been an increase in demand for young girls by both female and male paedophiles.

Most children involved come from marginalised communities in Sri Lanka, where they live in very low-income households. Such children are approached directly by sex tourists as well as indirectly through an agent, who is often an older boy no longer wanted for sexual gratification but who seeks younger boys in return for financial rewards. According to MPCP (1996):

> In the Sri Lankan context to use the term 'prostitute' to describe the occupation of these children is misleading, for they are 'consenting partners' to a homosexual relationship and willing to lend themselves to sexual abuse for money and goods, promises of jobs and a life abroad with their lovers.

However, for the majority of the children involved such promises do not come true. Many children suffer mental and psychological trauma while the rate of sexually transmitted diseases, including AIDS, has increased dramatically during the 1990s. Although as noted above, the extent to which abused children go on to act as pimps themselves indicates the potential for the cycle of abuse to continue.

Prostitution and child-trafficking has also grown substantially in Vietnam since the country has opened itself up to the West and introduced market reforms. According to Daniel (1996) at least 1,800 children had been trafficked into Cambodia in the previous year, but the figures were probably much higher: 'Parents are selling their children for 300, sometimes 400 dollars. Often, they think they are sending them to work in restaurants or as maids for a few months. They rarely see them again.' In Cambodia, children fetch high prices as the market for Vietnamese girls has been fuelled by a fear of AIDS in Thai sex workers. Little is being done to stop the trafficking in Cambodia. Indeed, some reports state that in Phnom Penh, many brothels are owned by the police with the police arranging dummy raids on their own brothels (Daniel, 1996).

According to Kristof (1996), several factors seem to be aggravating the problem of child prostitution in Asia:

> Rising economic development, which initially seems to increase the appetite for children more quickly than it reduces the supply; the rise of capitalism in places like China and Indochina, so markets emerge not just for rice and pork but also for virgin girls.

A Chinese superstition holds that sex with a virgin helps make a man young again, or that it can cure venereal disease; and perhaps very significantly, the fear of AIDS appears to be driving customers to younger girls and boys who are regarded as more likely to be disease-free. Of course, the children may in fact, be greater risks because of their youth: their vaginas and anuses are easily torn, creating sores and bleeding that permit STDs and the AIDS virus to spread. However, Kristof (1996) also noted that there was seemingly an increased desire for control of young girls by sex tourists, quoting Tisay, a 14-year-old streetwalker in Quezon City in the Philippines:

> Most men I know want younger girls, the younger the better, because then they can scare the hell out of us. Older girls can set a price, can set conditions, but younger girls can't do that … I'd prefer safe sex. But it's hard to insist that a man wear a condom. I'm small and I'm alone and I can't do anything about it if he doesn't want to.

Undoubtedly there are numerous facets to the increase in sexual demand for children. It may be possible that media and institutional awareness of the problem has increased the rate of reporting. Such a perspective is generous at best. In an earlier chapter on sex tourism in south-east Asia, one of the authors (Hall, 1992) pondered as to whether increased economic development in the region would lead to a lessening in the rate of tourism-related prostitution as economic conditions in the peripheral areas which were the major source regions improved. To an extent it has. However, there has also been a shift in the location of supply. In the same way that business corporations have shifted their manufacturing operations to take advantage of low labour costs so the procurers of women and children for the brothels of Asia have now shifted their attention to the poor areas of Cambodia, Vietnam, Laos, northern Thailand and southern China. In addition, many of these women now end up in Japan and the United States as the international trafficking in women continues to grow. Many of these, often forced, sex workers do not cater for the sex tourist, they work in brothels that are designed to cater for domestic demand. Nevertheless, sex tourism remains implicated in the institutional structures which allow such trafficking to occur, particularly because of the large amounts of wealth that tourist's foreign currency can bring to brothel owners and corrupt or ineffective officials while tourist expenditure sustains wider host attitudes towards women as commodities. Sex tourism is implicitly a component of the demand and supply for this body trade where the commodity is a physical and social being who has been reduced to an economic instrument.

This is also clearly not to suggest that there is not substantial opposition to trafficking in women and children. For example, in 1998, President Clinton identified trafficking in women and girls as a 'fundamental human rights violation' and sought to further develop and coordinate government policy on this issue including participation in the negotiation of a protocol on trafficking supplementing the Convention against Transnational Organized Crime; implementation of foreign aid programmes designed to prevent trafficking, assist victims, and prosecute traffickers; and consideration of legislation in the US Congress against trafficking in persons (Ralph, 2000). Most significantly being the bill (H.R. 1238) introduced by Senator Paul Wellstone in November 1999 'to combat the crime of international trafficking and to protect the rights of victims' (see p. 119). Moreover, a number of countries have also introduced similar legislation dealing with trafficking and sex tourism, including Australia, where child sex tourism has had a high media profile through the activities of ECPAT (Hall, 1998).

In December 1990, following pressure from ECPAT and members of the Australian Council for Overseas Aid (ACFOA) such as World Vision and UNICEF, Australia ratified the United Nations Convention on the Rights of the Child with the Convention coming into force in January 1991 (House of Representatives Standing Committee on Legal and Constitutional Affairs [HRSCLCA], 1994). Under Article 34 of the Convention:

> States Parties undertake to protect the child from all forms of sexual exploitation and sexual abuse. For these purposes, States Parties shall in particular take all appropriate national bilateral and multilateral measures to prevent:
>
> (a) The inducement or coercion of a child to engage in any unlawful sexual activity;
> (b) The exploitative use of children in prostitution or other unlawful sexual practices;
> (c) The exploitative use of children in pornographic performances and materials.

The obligations set in place by the Convention and the ongoing campaign by ECPAT and ACFOA for the Australian government to be active in the control of child prostitution in Asia were lent support by the activities of the Australian Human Rights Commissioner and the visit of Vitat Muntabhorn, the Special Rapporteur of the United Nations Commission of the Program for the Prevention of the Sale of Children, Child Prostitution and Child Pornography in 1992. Professor Muntabhorn proposed that the Australian government considered the possibility of extending national jurisdiction to cover the involvement of Australians overseas in 'transnational sexual exploitation (Muntarbhorn, 1993: para 106).

As a result of this public pressure the Australian state and federal government agreed to support such measures and on 5 July 1994 The Crimes (Child Sex Tourism) Amendment Act 1994 (Commonwealth of Australia Gazette, 1994: 1479). It inserted a new Part IIIA into the Crimes Act 1914 to deal with the activities of:

- Australians who travel overseas for the sexual exploitation of child prostitutes;
- those responsible for organizing overseas tours for the purpose of engaging in sexual relations or activities with minors; and
- those who otherwise profit from child sexual exploitation.

The Act makes such activities the subject of criminal offences punishable in Australia and sets out the following:

- the nature of offences committed under the new Part IIIA;
- defences to these offences;
- the provision of evidence by video link in proceedings under this Part; and
- the provision of rules relating to the conduct of trials under this Part.

The Act makes special arrangements for the protection of children giving evidence. If the attendance of the child witness at the court to give the evidence would cause the child psychological harm or unreasonable distress or subject the child to intimidation or distress, the court can direct that the child's evidence be given by video link. The court may also order payment of expenses incurred in connection with giving evidence in this manner. However, although the Act received general bi-partisan support, several sections of the Act attracted controversy (Hall, 1998). According to the Minister for Justice, Hon. Duncan Kerr, in his second reading speech relating to the bill:

> The principal aim of this legislation is to provide a real, and enforceable deterrent to the sexual abuse of children outside Australia by Australian citizens and residents … The Bill also focuses on the activities of those who promote, organize and profit from child sex tourism. Provided they operate

> from Australia, or have a relevant link with Australia, they, too, will be able to be prosecuted for their contribution to the abuse of foreign children …
>
> The Bill aims to achieve these ends by creating sexual offences, relating to conduct outside Australia, which will be punishable in Australia, and offences of encouraging or benefiting from child sex tourism, which may be committed in or out of Australia, and will be punishable in Australia provided there is a relevant link with this country. All these offences will have substantial penalties, ranging from 10 to 17 years imprisonment, or correspondingly high pecuniary penalties if a company is involved.
>
> (1994: 1, 2)

From a legal perspective the Act was significant in Australia in terms not only of the subject matter of the law but also in applying Australian law extraterritorially (Hall, 1998). In March 1996 the first Australian charged under the Act, Anthony Carr who sexually abused children in the Philippines, was convicted and sentenced (ECPAT Australia, 1996a, b). However, the evidence for this apparently was discovered when the alleged offender was arrested for child-sex offences allegedly committed in Australia, rather than as a result of any investigation directed at the overseas activity (Parliamentary Joint Committee on the National Crime Authority (PJCNCA) 1995).

In evidence presented to the PJCNCA's Inquiry into Organised Criminal Paedophile Activity in Australia some witnesses expressed concern that the legislation would not be seen as credible and therefore have a deterrent effect, until a conviction occurs under it (e.g. World Vision Australia, Evidence: 47 in PJCNCA, 1995). Similarly, the Australian Federal Police Association (AFPA) argued that unless there are some significant successful prosecutions under the legislation, it runs the risk of eventually being seen only as 'posturing' (Submission from the Australian Federal Police Association, 31 March 1995: 7, in PJCNCA, 1995: para. 4.55). Indeed, the AFPA argued that there is no evidence on whether the legislation has had a deterrent effect. In contrast, the Minister for Justice, Hon. D. Kerr, and the Australian Federal Police (AFP) both indicated to the Committee that they believed that the legislation was having some deterrent effect (Evidence, 149–50 in PJCNCA 1995: para. 4.54), although whether this effect had been more on the casual 'sex tourist' than the determined paedophile was less easy to determine. As AFP Detective Superintendent Paul Kirby stated: 'I'm inclined to think that the real paedophiles are just being more cautious about their activities and going to ground … I think the act has probably had a real effect on the casual sex tourist' (*Sydney Morning Herald*, 1995: 17).

Despite questions as to the efficacy of the Act as a deterrent to sex tourism, the Act and surrounding publicity has had considerable media and political profile in Australia with it serving as one of the cornerstones of Australia's National Plan of Action to Combat the Commercial Exploitation of Children and ECPAT's own campaign efforts (ECPAT Australia, 2000). For example, the PJCNCA (1995) noted the considerable efforts that have been made to publicise the new legislation, particularly to Australian residents departing for overseas through the distribution of leaflets by both ECPAT and the Australian Customs Service. Furthermore, in the diplomatic arena the leader of the Australian government delegation to the World Congress Against the Commercial Sexual Exploitation of Children, the Australian Ambassador to Sweden, noted the Australian efforts with respect to the adoption of extra-territorial laws, and trafficking in children and child pornography (Muntarbhorn, 1996).

The issue of paedophile activity, child pornography in terms of both sex tourism and virtual tourism on the Internet has now become a significant issue in Australian politics. For example, the Australian Department of Foreign Affairs and Trade (DFAT) conducted an inquiry into paedophilia amongst Australian diplomats (ECPAT, 1996c), while the New South Wales (NSW) Royal Commission into Police Corruption revealed details of paedophile activity, police corruption, and police lack of action and mismanagement of cases (ECPAT, 1996d). In the case of the NSW Royal Commission it became apparent that there was an awareness of paedophile activity but that it was typically not regarded as a serious crime and allegations, particularly involving institutions such as the Catholic Church, were often not acted upon. Similarly, with respect to the passage of the Child Sex Tourism Act, Mrs Easson, a member of the House of Representatives stated: 'We have all known for years that the child prostitution racket is very widespread in Asia and that it exists in Australia' (1994: 146) leading Hall (1998: 93), in a study of the development and policy implications of the act to comment, 'This begs the question, why has government taken so long to act given that sex tourism has been an issue for at least 20 years?'

This chapter has discussed issues of sexual slavery and the trafficking of women and children. These issues are, for many people, well beyond the margins within which sex tourism may be acceptable. In the forced supply of sexual services the sex tourist has typically moved into the illicit. Although condemned by many institutions, the trafficking of women and children for the supply of such services appears to be continuing to grow if the evidence not only from Asia, but also from Europe and North America is to be believed. Yet there are also dangers in placing child prostitution in a categorisation removed from other forms of sex tourism. By distinguishing it, it assumes abnormality on the part of the sex tourism, while the fact may be that such practices are equally bound within wider social contexts. For a commentator like O'Connell Davidson (2000: 69), such drawing of boundaries fails to note that 'the majority of sex tourists who use child prostitutes are first and foremost *prostitute users* who become child sexual abusers through their prostitute use, rather than the other way about'. Child prostitute users are, in her view, therefore not manifesting abnormality, but are 'expressing attitudes and desires that are very ordinary and widely accepted' among tourists, that is, an expression of power and racism 'routinely voiced in mainstream white Western culture' (ibid.).

Certainly the trafficking of women and children is not confined to the sex industry. Indeed, the sex industry represents but part of a wider global exploitation of those vulnerable through poverty, age and gender. For example, in March 2000 the New Zealand Department of Labour seized the assets of Wiliwan Sivoravong. Sivoravong ran the companies KC Fashions and Sivoravong Fashions in Auckland. In the evidence one Thai worker, Panja Hirikokul, told of how she was promised 35,000 baht per month plus free accommodation for working at Sivoravong's factory in Auckland over six days a week. In spite of warnings from the New Zealand authorities she travelled to Auckland in February 1997. There she ended working for thirteen hours a day for NZ$4 per hour, but received no pay for the first six months to offset 'loans' for visa and travel costs. Her passport was withheld from her. In October 1997 her leg became badly swollen, but her employer still insisted that she should continue working. Other women were subsequently found to have similar stories, having been recruited by Sivoravong's mother in Thailand (*New Zealand Herald*, 2000: 1).

Again, in March 2000, the New Zealand police, as part of an international police action arrested what were described as 'three key players' (Mager and Andrews, 2000) who had been involved in immigrant smuggling. Detective Sergeant Craig Turley was quoted as saying that those who could not pay for altered and forged documentation were forced into a life of crime. 'The syndicate is living off these people. It's a multimillion-dollar business ... We believe some [debtors] have been forced to smuggle drugs between countries' (Mager and Andrews, 2000).

Such stories are repeated time and yet time again. Small children are forced to labour long hours that eventually destroy their health; men, women and children become but discardable cheap units of production. Often, the purveyors of this human cargo are their fellows, living in the same villages, who know but one way to escape the poverty that exists, and that is by brutalising others. Attempts to control the problem of trafficking in areas of both supply and demand appear to have had little success, the commoditisation of the body for the gendered economy of sexual slavery, prostitution and factory labour appears to be more than a match for governments and agencies, and it is to this vexed question of control that the next chapter will turn.

Finally, in Chapter 3 reference was made to 'virtual sex tourism' on the Internet. Hughes (1999) argues that Internet portrayal of prostitution, pornography, the 'ordering' of mail brides through the Internet – all of these constitute yet another form of trafficking. She writes: 'The European Union defines trafficking as a form of organized crime. It should be treated the same way on the Internet. All forms of sexual exploitation should be recognized as forms of violence against women and human rights violations, and governments should act accordingly' (Hughes, 1999: 182). Such initiatives raise significant questions of definition which determine policies that seek to balance a need to protect the exploited with rights of individuals to access non-exploitative text and images. The debate about trafficking thus here spills over into issues of definitions of pornography, depraving and corrupting, thereby again illustrating how issues related to sex tourism are securely embedded within broader social issues, power structures and *mores*.

Key points: Bill (H.R. 1238) to combat the crime of international trafficking and to protect the rights of victims.

Directs the Secretary of State to: (1) report annually, with the assistance of the Task Force, to Congress describing the status of international trafficking; and (2) ensure that U.S. missions abroad maintain a consistent reporting standard and thoroughly investigate reports of trafficking. Requires U.S. mission personnel to seek out and maintain contacts with human rights and other nongovernmental organizations, including receiving reports and updates from such organizations and, when appropriate, investigating such reports.

Makes any government of a foreign country identified in the latest report as one that has failed to take effective action towards ending the participation of its officials in trafficking and that has failed to investigate and meaningfully prosecute those officials found to be involved, ineligible for police assistance, subject to a presidential waiver if in the U.S. national interest.

Amends the Immigration and Nationality Act to provide for a non-immigrant classification for trafficking victims. Provides for a waiver of grounds for ineligibility for admission with respect to such an individual if the Attorney General considers it to be in the national interest to do so.

Directs the Attorney General and the Secretary of State to jointly promulgate trafficking regulations for law enforcement personnel, immigration officials, and Foreign Service officers concerning response training and victim treatment and protection.

Authorizes: (1) the Secretary of Health and Human Services (HHS) to provide, through the Office of Refugee Resettlement, assistance to trafficking victims and their children in the United States; and (2) the President to provide programs and activities to assist trafficking victims and their children abroad.

Authorizes appropriations for the Task Force, the Secretary of HHS, and the President. Bars the use of funds made available to carry out this Act for the procurement of weapons or ammunition.

7 The role of the state

Regulating sex and travel

This book has argued that sex tourism occupies marginal spaces that are in themselves a component of the inherent marginality of tourism in contemporary Western society. The previous chapter discussed how sex tourism may well be illicit, but, as earlier chapters noted, the inter-relationship between sex tourist and sex worker may also move from the symbolic to the pragmatic and thence progress to the functional. The marginal spaces of sex tourism are therefore consistently shifting according to different social, cultural, economic and political factors. Indeed, the contested, multi-layered, nature of sex tourism implicitly suggests that this, at times, transitory space of inversion, provides the capacity for new understandings, perceptions, relationships and representations of sexuality, travel and the sex industry. Due to its increasing visibility, sex tourism therefore possesses the potential to act as a catalyst for encapsulating or creating social change, and in Australia and New Zealand, the countries with which the authors are most familiar, this has certainly become evident with respect to changing attitudes towards the sex industry. Such changes become codified through the regulatory actions of the state which has historically served to implement a moral code with respect to sexual behaviours through legal sanctions. Indeed, the role of the state with respect to sex tourism may not necessarily be one of prohibition or prevention. In some jurisdictions sex tourism has been implicitly or even overtly encouraged in order to attract foreign exchange and encourage economic development. This, to some, perhaps paradoxical relationship between sex tourism and the state, reflects the complex multilayered nature of understanding sex tourism to which we have often referred within this book.

This chapter discusses the nature of the interrelationship between the state and sex tourism particularly with respect to the 'control' of sex tourism. However, in examining this relationship we will also note the various intersections that exist with other considerations in the development of an understanding as well as the position of sex tourism within contemporary capitalism. The chapter first discusses various aspects of the interrelationship between sex tourism and economic development in South-east Asia, a region often featured in discussion of the development and control of sex tourism (e.g., Jeffreys, 1997). The chapter then discusses some of the issues that surround decriminalisation, legalisation and control, including problems with respect to controlling sex tourism in the communications age. The chapter then goes on to examine the way in which the state and contemporary capitalism effectively commodify the body in a manner which not only creates the spaces of marginality in which sex tourism exists but which also commodifies that space itself as a place to be consumed.

Sex, tourism and economic development: the case of South-east Asia

Sex tourism has been recognised as an overt component of the touristic attractiveness of several countries of South-east Asia since the late 1960s (e.g., see ISIS, 1979; O'Grady, 1981; Bishop and Robinson, 1998). In South-east Asia the institutionalisation of sex tourism occurred when the prostitution associated with American military bases and Japanese colonialism was transformed into a component of the international tourism industry and an integral component of national and regional economic development. Prostitution is technically illegal in many South-east Asian countries, but the law is poorly enforced. The prevailing sentiment appears to be 'that what a tourist does in the hotel room, is none of the authorities business' (Sentfleben, 1986: 22), particularly when it results in economic returns to a region and members, usually male, of the local authorities and business classes. Given the illegal and often casual nature of much sex tourism it is extremely difficult to determine the exact number of sex tourists to the region. According to one estimate from the mid-1980s, 'between 70 and 80 per cent of male tourists who travel from Japan, the United States, Australia, and Western Europe to Asia do so solely for the purpose of sexual entertainment' (Gay, 1985: 34). The present-day figure is unknown especially as there has been a massive growth in interregional travel and trade (Hall and Page, 2000). The changing nature of tourism in the region which now has greater emphasis on family-oriented resort tourism, marine and ecotourism has possibly meant that the relative number of sex tourists to other markets has declined but the number still remains significant (Oppermann, 1998; Hall and Page, 2000).

From a feminist perspective, the study of the sex tourism industry encapsulates many of the problems which are fundamental to women and development: 'Prostitution is both an indiction of an unjust social order and an institution that economically exploits women. But when economic power is defined as the causal variable, the sex dimensions of power usually remain unidentified and unchallenged' (Barry, 1984: 9). As Claire and Cottingham (1982: 208) report, often women who have fled rural poverty only to be forced into prostitution by urban unemployment, are 'victims of the double standard. Women who have been raped, jilted, or taken advantage of no longer fit the chaste wife-mother-sister ideal and are ostracized by nearly all sectors of society'. Historically, the economic and social problems of Asian women tend to be viewed by Western feminists as stemming from the patriachial nature of local cultures. For example,

Truong (1983) reported that 'in eastern societies, concubinage and brothels have existed for centuries and carried a distinct class connotation'. The influence of the patriarchial nature of society on the manner in which sex tourism comes to be defined as a 'problem' does not negate the substantial insights made into sex tourism by feminist researchers. However, researchers such as Ong (1985: 2) argued that 'the new commerce in the labor-power and bodies of Asian women, is more rooted in corporate strategies of profit maximisation than in the persistence of indigenous values'. From the perspective of political economy, sex tourism may be regarded as a result of shifts in the international division of labour within a globalised economy and the development of consumerism in Asian countries of which tourism is a major constituent. Such shifts may only serve to reinforce the gendered nature of local power relations in which some bodies assume the function of commodities to be consumed by tourists or locals. The role of tourism in the relationship between consumerism and globalisation is observed by Bishop and Robinson (1998: 108): 'Not only is shopping increasingly understood as the moral equivalent of sightseeing for the tourist, but the tourist locale – even the land itself – is represented as consumable goods.' Furthermore, to Bishop and Robinson 'With all tourist sites, commerce depends on the construction of a desirable other – often one that titillates as well as appeals – capable of attracting outsiders' (1998: 61). They then went on to argue:

> This construction can create inequitable interactions between local and traveler that actually serve to reinforce disparity while being represented as mutually beneficial. In these international interactions, including the sexual ones, the flow from center to periphery, from here to there, is virtually unidirectional; the trickle in the opposite direction largely provides education for an academic elite and political class. The disparity of interactions can be charted in this flow: when 'they' come 'here', we educate them; when 'we' go 'there', they service us.
>
> (1998: 61)

Despite their criticism of the tourism studies literature, the well-articulated critique of sex tourism by Bishop and Robinson which had integrated political economy and cultural studies traditions had been anticipated within the tourism field. For example, Graburn argued:

> The phenonemon of prostitution in the third world is particularly crucial because of the economic power differential between the buyer and the seller. Furthermore there is a direct analogy between prostitution in the Third World and that in the metropolitan resort centers where the prostitutes are disproportionately drawn from disadvantaged sections of the population who may have similar economic problems illuminating forms of 'internal colonialism' commonly found in stratified, industrial societies.
>
> (1983b: 442)

The flow of international capital and visitors to the less developed countries has led several commentators to conclude that tourism is prostitution (e.g., ISIS, 1979; Sousa, 1988) and an inevitable consequence of mass tourism. For example, Rogers stated:

> The nature and character of modern tourism which is strongly centered around [an] unquenchable [thirst for] profit and the sexual gratification of men from the First World cannot but breed and perpetuate the prostitution of deprived and dispossessed women and children of the Third World.
>
> (1989: 20)

More recently, Seabrook (1996: 167) argued that 'resistance to sex tourism ... should be part of a wider campaign against tourism in general ... The ideology of cheap holidays is part of the ideology of cheap goods, cheap labour and cheap sex.' Such sentiments would be anathema not only to many in the tourism industry, but such a perspective also indicates a sharp departure from much of tourism research which generally ignores questions of ideology, gender and the capitalist system within which tourism is situated and the often marginal nature of the tourist experience itself. However, as argued throughout the book, failure to recognise the marginal and multilayed spaces of tourism only serves to further marginalise many of the workers who occupy those spaces. If solutions are to be found to the vexed issues of sex tourism, it therefore becomes essential that debate and ideas be allowed to come into a wider public sphere in which they can be adequately interrogated rather than continue to be blindly ignored.

Tourism is a major industry in South-east Asia. For most of the countries in the region tourism is one of the most important sources of foreign exchange and employment generation. International tourism visitation has demonstrated almost continued growth since the late 1970s at a rate well above the world average. According to Bacani (1998) tourism accounts for around 10.3 per cent of Asia's GDP. Perhaps somewhat paradoxically, the 1997–98 Asian financial crisis has served to make tourism even more important for the region as countries seek to gain urgently needed foreign exchange and attempt to encourage renewed growth at a time of economic recovery (Hall and Page, 2000). Tourism has therefore been integral to the region's economic development. However, the commodification of the human body by the sex tourism industry cannot be explained simply by reference to state economic strategies. Instead, as the above discussion suggests, there are multifaceted and interrelated reasons for the development of sex tourism in the region.

One of the authors identified four stages in the development of sex tourism in an earlier account of sex tourism in the region: indigenous prostitution, economic colonialism and militarisation, substitution of international tourists for occupation forces and rapid economic development (Hall, 1992). The discussion below presents a revised account of sex tourism development with provision being made for the internationalisation of legal and political responses to sex tourism in the mid-1990s and the impacts of the Asian financial crisis in the late 1990s.

Indigenous prostitution

Prostitution in South-east Asia clearly existed before the arrival of tourists with 'domestic' prostitution continuing to constitute a major component of most countries' sex industries. Writers such as Truong (1990) and Hill (1993) have argued that Buddhism plays a major role in perpetuating the patriachial nature of Thai culture. This point is also reflected in the comments of Skrobanek (1996: vii) who commented that 'certainly there are elements in Thai society which contribute to the growth of commercialisation of human relationships, both in the family and community, and to the commodification of women's body and soul'. An issue raised by a number of other commentators. For example, Richter, in her seminal work on the politics of tourism in Asia argued, 'Perhaps because Thailand was never colonized, and also because the nation has a history of concubinage and prostitutes in its traditional culture, opposition was slow to recognize the difference in scale, violence, and social decay implied by sex tourism' (1989: 84). More recently, Seabrook (1996: 79–80) observed that 'the monastery, military and monarchy had degraded the social and cultural position of women to such a degree that it was easy for the market to do the rest'. Indeed, the importation of elements of Brahminical culture into Thai society has allowed concubinage and polygamy to be legitimised and has cast a ready-made framework within which sex tourism can be culturally acceptable. As O'Malley (1988: 107) argued, 'The result was an erotic industry promoted by a government hungry for foreign exchange and built upon the solid base of a hundred years of institutionalized prostitution.'

While social and religious institutions have had a significant role in commodifying women, Buddhism cannot itself be held solely responsible. As Bishop and Robinson (1998: 160) noted 'virtually all the world's major religions include patriarchal power structures that do not necessarily lead to the establishment of prostitution as a major industry ... despite rumours to the contrary, Buddhism explicitly prohibits the practice of prostitution.' Indeed, it is with respect to patriarchal power structures that common elements emerge between the various countries of the region that have engaged in sex tourism. For example, one of the ironies of the current Japanese involvement in sex tourism is that the Japanese used to export their own prostitutes *Kara-Yuki San* to their colonies. *Kara-Yuki San* were bonded Japanese women who were sent abroad to serve as prostitutes in ports frequented by Japanese merchants and soldiers (Hawkesworth, 1984; Matsui, 1987a, b). However, since the 1920s when the Japanese government issued the Overseas Prostitution Prohibition Order and with the prohibition of legal prostitution in Japan in 1958, women from the former colonies 'are now imported into Japan as prostitutes' (Graburn, 1983b: 440; also see Leheny, 1995).

As previously noted, such was the significance of prostitution in Japan's male-dominated culture that following both the 1911 and 1923 earthquakes the reconstruction of licensed prostitution quarters took priority over the rebuilding of schools (Robins-Mowry, 1983). Moreover, domestic prostitution may not only be based on gendered power structures but also on race and culture. For example, in Taiwan the majority of prostitutes are not Han Chinese but instead come from the island's aboriginal population which is Polynesian-Malayan in origin and which lives in the marginal rural areas (Senftleben, 1986). Similarly, many of the minority ethnic groups in Myanmar and Thailand are often disproportionately represented in prostitution in relation to the percentage of the general population they represent. Therefore, the factors which lead women and men into domestic prostitution may be regarded as an inter-related series of marginalities relating to gender, race and economics, and it was upon this set of unequal power relations that the second stage of sex tourism development in the region was built.

Economic colonialism and militarisation

The second stage is that of economic colonialism and militarisation in which prostitution is a formalised mechanism of dominance and a means of meeting the sexual needs of occupying military forces (e.g., Enloe 1990, 1992). In this stage the occupied culture's general acceptance of various forms of prostitution has been used as a justification for economic or military enforced prostitution or, as in the case of Japanese militarism in the 1930s and 1940s, was used as a means of exercising power on host populations. In addition, this stage commences the economic dependency of certain sections of host societies on the selling of sexual services as a means of economic growth and development. For example, in the case of Taiwan, hot spring resorts which provided for the spatial concentration of tourist-related prostitution activity were first developed under the Japanese colonial era between 1895 and 1945. According to Robins-Mowry:

> the Japanese Government organised a system to service the Occupation troops, systematically recruiting women – patriotically – to serve as prostitutes. From their own experience on mainland China, the authorities considered this essential for any military occupation. The women were recruited, according to Morosawa Yoko, as the 'breakwater to protect Japanese women's chastity' – meaning, of course, the chastity of less needy daughters and wives.
>
> (1983: 248)

However, military-related prostitution continued after the end of Japanese military occupation with the resort of Peitou achieving rapid development in the post-war era with legalised prostitution until 1979 as a favoured rest and recreation area for American forces (Senftleben, 1986).

Militarisation also played a major role in the development of the Filipino sex tourism industry. While the United States retained a military presence in the Philippines, the 12,000 registered and 8,000 unregistered hostesses in Olongapo City provided the major source of sexual entertainment for the US military personnel based at Subic Naval Base and Clark Air Force Base. The City was economically dependent on the military presence and a number of city ordinances were written which allowed prostitution to be legitimised, regulated and protected by the local state (Moselina, 1979; Philippine Women's Research Collective, 1985). For example, the city enforced an anti-streetwalking ordinance which ensured that the soliciting of customers could only occur inside clubs thereby assuring club owners of fees derived from the provision of sexual services (Claire and Cottingham, 1982). The American military presence was also formalised in Thailand when in 1967 the Thai government signed an agreement with the US government to provide Rest and Recreation facilities in Thailand for American troops in Vietnam (Bishop and Robinson, 1998).

The American military presence in South-east Asia, and in Thailand in particular, created the foundation for sex tourism in several ways. First, it maintained and reinforced indigenous power relationships which were exploited through prostitution. Second, it served to commodify local bodies for the pleasures of foreigners and thereby increase the market value of female sexual capacity. Third, it created a series of economic structures and dependencies which would be filled by the international tourist once the military forces departed.

International tourism

The third stage was marked by the substitution of international tourists for occupation forces. Following periods of occupation and the restructuring of traditional economies within the post-war international economic order, sex tourism became a formal mechanism for obtaining foreign exchange and of national development. A common element in this third stage is the authoritarian nature of governments during periods in which sex tourism was being promoted by government and the tourism industry. For example, the authoritarian nature of successive South Korean governments through the 1970s to the late 1980s played a major role in the commoditisation of women through *kisaeng* tourism. Prospective *kisaeng* endured lectures by male university professors on the crucial role of tourism in the South Korean economy before obtaining their prostitution licences. At the government 'orientation programme' for sex workers, women were told 'Your carnal conversations with foreign tourists do not prostitute either yourself or the nation, but express your heroic patriotism' (Gay, 1985: 34). Perhaps more telling is the report of the South Korean Minister for Education who stated that 'the sincerity of girls who have contributed with their cunts to their fatherland's economic development is indeed praiseworthy' (Witness 2, 1976 International Tribunal on Crimes against Women, 1976: 178 in Symanski, 1981: 99).

It is possible that the denial of individual rights by authoritarian regimes may have encouraged the perspective that individuals are sexual commodities to be utilised for furthering the national economic good. Similarly, the Thai government placed great emphasis on the promotion of R&R and tourism in the economic development of Thailand. For instance, in 1980 Booncha Rajanasthian, Thailand's Vice Premier, asked all provincial governments 'to consider ... forms of entertainment that some of you might consider disgusting and shameful, because we have to consider the jobs that will be created' (cited in Bishop and Robinson, 1998: 10).

As part of the development of international tourism not only were bodies incorporated into the international tourist economy but also the culture. For example, 'The stereotype of Thailand as the playground of the Western world dominates the public imagination outside the country and is continually reiterated in the popular media' (Bishop and Robinson, 1998: 16). Indeed, advertising for many of the South-east Asian nations openly plays on the notion of the 'exotic orient', South Sea romanticism and the image of a 'lost paradise' which has existed since the seventeenth and eighteenth centuries (Selwyn, 1993; Douglas and Douglas, 1996). For instance, Davidson (1985: 18) reported a Frankfurt advertisement which stated, 'Asian women are without desire for emancipation, but full of warm sensuality and the softness of velvet', characteristics which are related too in more contemporary tourism advertising. For instance, Singapore Airlines proclaims 'Singapore Girl you're a great way to fly' along with soft images of airline hostesses waiting to provide service for customers.

Rapid economic development and international controls

The fourth stage of sex tourism for most of the nations of the region was that of rapid economic development. As the author commented in 1992 (Hall, 1992): 'It is as yet unknown whether increased standards of living will reduce dependency on sex tourism or whether the growth of consumerism will become a new factor in the maintenance of the sex tourism industry.' Indeed, in 1976 the head of the Tourist Authority of Thailand commented that 'prostitution exists mainly because of the state of our economy. If we can create jobs, we can provide per capita income and do away with prostitution' (quoted in Truong, 1990: 179). However, while economic growth was undoubtedly rapid from the 1970s through to the mid-1990s it was also uneven. For example, the regions of the northeast and northern provinces of Thailand along with displaced minorities from Burma on the Thai–Burmese border have continued to provide a major source for child and female prostitutes. The economic marginality of the regions forces many rural households to depend on the remittances provided by migrant girls. The northern Thai provinces have continued to remain structurally disadvantaged within the Thai economy with much of the export and large-scale tourist growth concentrated in the southern and central provinces and 'if no new income sources are created by the typical prostitutes' earnings; a vested interest and dependency upon the continuation of the sex industry is created in the rural hinterland' (O'Malley, 1988: 110). Therefore, given the lack of economic development in the north, there would appear to be an assured supply of workers for the sex industry based in the nation's urban and industrial centres. As Bishop and Robinson argued:

> It is hard to see how economic growth arising from prostitution-based tourism could do away with prostitution. On the contrary, since the market sets up a permanent demand for a sex and age-specific labor force which, as it happens, ages very rapidly, the way to assure the constant availability of fresh supplies from the rural areas is precisely to pursue national planning policies that systematically de-emphasize agriculture and displace fishing and to withhold resources where they historically constituted the economic base.
>
> (1998: 99)

Nevertheless, some controls and limiting factors on prostitution have been put in place. Most significantly, AIDS has become a major source of concern to governments in the region. For example, in 1989 the Thai Public Health Ministry actively started campaigning against prostitution and the promotion of Thailand as a sex tour destination. The primary reason for the campaign was the recognition that sexually transmitted diseases such as AIDS could pose major problems for Thailand's rapidly growing tourism industry and for the Thai economy in general. The Asian Development Bank calculated that 'through the death and disablement of AIDS sufferers – usually from the economically productive 20–40 age group – Thailand ... had lost almost $3 billion so far and by 2000 this would rise to $3.5 billion a year if the disease went unchecked' (quoted in Hall, 1996). The concern for the manner in which the AIDS dimension of sex tourism was seen to be harming the country's tourism industry is well indicated in the comments of the Thai Deputy Public Health Minister, Suthas Ngernmuen:

> Thailand's profitable tourist industry has been an inhibiting factor in promoting AIDS awareness ... More than two-thirds of the overseas visitors entering Thailand are single men, and medical officials avoided publicising the appalling AID statistics for fear of damaging the country's healthy tourist business ... But it is long past time for the government to change Thailand's image as a sexual paradise.
>
> We should promote tourism in more appropriate ways, and campaign more against AIDS.
>
> (in Robinson, 1989: 11)

Sex tourism therefore represents a major dilemma for the Thai authorities. Sex tourism continues to be a major tourist attraction and hence a source of foreign currency but authorities are increasingly worried by Thailand's reputation as the sex capital of Asia which may repel other potential markets. For these reasons the Thai government created the 'Visit Thailand year ' in 1987 in order to refurbish its image and to de-emphasise sex as an attraction (Cohen, 1988). Nevertheless, as economic growth continues to be uneven, peripheral rural areas will continue to furnish Thailand's sex industry with its raw material.

As well as attempting to change image and marketing campaigns, countries in the region have also begun to work together in order to stop child prostitution. Led primarily by ECPAT (End Child Prostitution in Asian Tourism), an international campaign has successfully managed to get child sex tourism legislation enacted in a number of countries in the region and in Europe (Hall, 1998). For example, Australia enacted the Crimes (Child Sex Tourism) Amendment Act 1994 to deal with the activities of Australians who travel overseas for the sexual exploitation of child prostitutes; those responsible for organising overseas tours for the purpose of engaging in sexual relations or activities with minors; and those who otherwise profit from child sexual exploitation. As Hall (1998) argued, ECPAT has been able to create an awareness among both politicians and the public about child sex tourism in South-east Asia. Such was the success of the ECPAT campaign in creating moral indignation and therefore political action that it almost passed by without notice that the term 'sex tourism' is not even defined in the Act. Undoubtedly, ethical concerns over undesirable social impacts of international travel were at the forefront of debates over the Act. Unfortunately, such 'awareness' of the negative impacts of tourism has been restricted to a narrow range of Asian concerns. Little concern was expressed in discussion of the legislation about the sexual activities of international visitors to Australia and the various campaigns of the Australian Tourist Commission which sought to display bikini-clad women in order to create a favourable and attractive image to certain market segments. In this situation, it may therefore be argued that while something has been seen to be done in the solution of 'sex tourism' issues, in reality, very little fundamental change has occurred (Hall, 1998).

The impact of the Asian financial crisis

The contemporary stage of sex tourism development in South-east Asia is one marked by the impacts of the financial crisis which began in mid-1997. As noted above, tourism has become even more important to the various countries in the region as a source of foreign exchange. As the previous chapter noted, in this climate the commodified body, usually a woman's, remains an important source of income generation through sex tourism and trafficking. Governments therefore implicitly, and occassionally explicitly, exploit bodies as a natural resource in much the same way as the rainforests of the region are sold and raped. Indeed, the opening up of Vietnam, Laos, Cambodia, southern China and Myanmar to tourism and the effects of the financial crisis has only sought to expand the range of low-cost sex tourism opportunities for visitors to the region. In this setting sex tourism to the region remains as problematic as when it first began to be discussed in the media in the late 1970s.

The political economy of sex tourism

Sex tourism is integral to the economic base of several regions of South-east Asia. The Philippine Women's Research Collective (1985: 36) argue that the 'insidious tourist first attitude' of many governments and their advisers has transformed the processes of development to place the economic 'needs' of a modernising economy well ahead of wider socio-economic concerns which feed into the growth of tourism-related prostitution such as rural migration (Bishop and Robinson, 1998). More critically, Enloe succinctly observed:

> Sex tourism requires Third World women to be economically desperate enough to enter prostitution; having done so it is made difficult to leave. The other side of the equation requires men from affluent societies to imagine certain women, usually women of color, to be more available and submissive than the women in their own countries. Finally, the industry depends on an alliance between local governments in search of foreign currency and local and foreign businessmen willing to invest in sexualized travel.
>
> (1990: 36–37)

If Enloe's analysis is correct, and much of the argument in this chapter would agree that it is, then how do we control sex tourism? Fish (1984a, b) has argued that an effective means of controlling sex tourism would focus on placing a heavier proportion of the costs of law enforcement and sanctions on the hotels, agencies and their customers. However, such an approach built on the assumptions of schedules of the elasticity of demand fails to recognise the broader economic, political and socio-cultural context within which sex tourism occurs. Short-term measures such as counselling may be useful, but sex tourism demands long-term solutions. 'Banning prostitution may be counterproductive and only create even greater hardship for the already impoverished women who engage in it' (Shirkie, 1982: 6). Similarly, Truong (1983: 534) argued that 'legislation to protect prostitutes, and to improve their working conditions and occupational health, is preferable to legislation that would deprive them of their livelihood'.

The issue of the right to be a prostitute remains extremely controversial and is an example of the problem of competing rights which has exercised critics of the efficacy of rights approaches. For example, Klerk (1995: 16) argues, 'It would be a violation of the principle of self-determination of individuals to forbid to prostitute. Moreover, a prostitute has the right to let another person exploit her, and she might have good reasons for that.' Therefore, from Klerk's perspective, it is 'not logical to restrict the concept of traffic to prostitution', when it accepts prostitution as a 'normal job' (1995: 17). Indeed, a very strong case exists for prostitution to be professionalised. Lap-Chew states:

> The more 'professional' the sex worker, the more care she takes of herself. The more 'legal' or 'legitimate' she feels, the less she will be afraid to report abuse and exploitation, the more she will seek health and other kinds of care for herself and be able to develop a degree of professionalism in her work. Is this not an argument in favour of recognition of prostitution as a form of legitimate work?
>
> (1995: 2)

In contrast, Barry, in writing up the final report of the International Meetings of Experts on Sexual Exploitation, Violence and Prostitution argues, 'Clearly prostitution cannot exist as a right because it negates already established human rights of the prostitute woman to human dignity, bodily integrity, physical and mental well-being' (UNESCO and Coalition Against Trafficking in Women, 1992: 6). Similarly, Jeffreys (1997: 319) observes that the establishment of the right to prostitute, 'transforms the right of some men to abuse women in prostitution and of others to make a profit from that use – interests which arguably provide the real fuel for the pro-prostitution position – into a woman's rights to have her human rights violated'. In seeking to argue that there is little sense in separating trafficking from prostitution, she then goes on to cite Raymond (1995: 2) who states, 'Prostitution, of course, is the goal of sex trafficking and builds the base for the trafficking in women and children … When prostitution is accepted by a society, sex trafficking and sex tourism inevitably follow.'

The authors do not accept the position of Jeffreys (1997: 331) in quoting Santos, that:

> those who accept that prostitution is 'an inevitable social institution' accept that, sex, however it is obtained, either by coercion, commercialization or even seduction, is a male right, and that bodies of women and children, and men too, can be and should be packaged and sold as a commodity because there is a buyer and there is a seller.

As Fraser and Nicolson observed:

> To construct a universalistic social theory is to risk projecting the socially dominant conjunctions and dispersions of her own society onto others, thereby distorting important features of both. Social theorists would do better first to construct genealogies of the categories of sexuality, reproduction and mothering before assuming their universal significance.
>
> (quoted in Yeatman, 1990: 291, in Jeffries, 1997: 340)

Instead, the authors follow the approach of Anti-Slavery International (ASI) (1995) which:

> believes that the definition of prostitution as commercial sex work has more scope than an exclusively abolitionist approach for enhancing the welfare of women and men whose sexual services are sold. By looking at commercial sex as work, in labour conditions, those involved can be included and protected under the existing instruments which aim to protect all workers and, where appropriate, forced labour and migrant workers; all persons from violence; and women from discrimination.

The approach of ASI focuses substantial attention on issues of child and forced prostitution but recognises the right of individuals to engage in sex work if they so wish. It is the argument of the authors that prostitution needs to be legalised so that the conditions under which the sex worker operates and the sex industry is run are as transparent as possible to external evaluation. The illicit marginal spaces need to be brought out of the shadows if they are to be effectively controlled. To deny the concept of sex work as being a professional service, as having like other jobs its good or bad days, is to continue to marginalise and stigmatise. This short-term measure would hopefully improve the health and economic conditions of prostitutes at a minimum. However, in the long term it may also lead to greater transparency of the web of institutionalised exploitation within which trafficking does, and tourism may, operate and reinforce in some circumstances, such as those that exist in parts of South-east Asia. Campaigns, demonstrations and rallies against sex tourism are likely only to lead to superficial changes to the sex industry. For example, ECPAT's campaign against child sex tourism in Australia, noted above, has had only a marginal impact on creating a broader understanding of sex tourism issues in Australia (Hall, 1998).

One cannot condone sex tourism, especially child sex tourism, if it is coercive in nature. However, it is also important that decisions and policies that are formulated with respect to sex tourism are given serious and considered debate. In the case of sex tourism in Australia this did not happen. The Australian Child Sex Tourism Bill was able to pass because it focused on a small area of sex tourism policy on which moral, and to a lesser extent, ideological consensus could be reached. Such was the success of the ECPAT campaign in creating moral indignation and therefore political action that, as previously noted, it almost passed by without notice that the term 'sex tourism' is not even defined in the Act. As Mr Slipper, member for Fisher, noted 'that the bill should more properly be entitled the Crime (Overseas Exploitation of Children) Bill … I think that name is a more appropriate name for the bill. The bill's title at the moment tends to be emotive, but I suppose that is a matter not of substance' (House of Representatives, *Hansard*, 29 June 1994: 2345). Child sex tourism is an important issue, but the numbers of children engaged in prostitution because of tourism is nothing like that of those sex workers who are above the age of consent in the countries in which they work. As the PJCNCA observed:

> Most sexual offences against children are committed by their relatives and neighbours who are not paedophiles in the strict sense of the term and who do not operate in any organised or networked way ... There is no evidence to suggest that organised paedophile groups have ever resembled what are traditionally thought of as 'organised crime' groups in size, aims, structures, methods, longevity and so forth ... There is no evidence of any current organised promotion or arrangement of tours by Australian paedophiles to overseas destinations known to be attractive to them. However, informal networking among paedophiles may assist some tourists going overseas to commit paedophile offences.
>
> (1995: para. 3.85)

Tourism is clearly an element in the reason for prostitution occurring in certain locations in South-east Asia and Australasia, but so also are gender, culture, the pattern of economic development, racism, poverty and wealth distribution, highly patriarchal societies and material interests. In examining media reports and, more particularly, government reports and parliamentary debates on the child sex tourism debate in Australia there is often an impression given that 'something has been done'. Perhaps it has. Nevertheless, because of the failure to closely examine the marginal space which sex tourism occupies, broader questions are being left unasked and extremely significant issues, such as gender, economic and power relations, are typically relegated to academic rather than policy discussions. As Hall (1998) argued:

> Governments of countries, such as Australia, while taking action on child sex tourism, fail to recognize that in their own tourism advertising, they also promote the commodification and objectification of the sexual body. But, of course, the portrayal of available female bodies on Bondi or Bali has absolutely nothing to do with child sex tourism ... *does it?*

This failure to acknowledge the embodied nature of tourism is highly significant. The tourism industry 'rests on the physical display of bodies perceived as fundamentally, radically, different from those of the majority of the audience who pays to see them ... The live presence of the performers is crucial to this economy of pleasure, for it provides the guarantee of the authenticity on which such commodification exists' (Desmond, 1999: 251). Sex tourism is an integral part of the commodification of body, culture and place on which the tourism industry is based. However, the commodification of sexuality is wider than just individuals. It also has to be seen in relation to places. Sites of seduction are created where the tourist and the investor are to be seduced. There may be 'red light' districts or they may be the alluring promises of place promotion. Either way tourism becomes not simply a provider and creator of demand for prostitution, but in the very processes of creating places attracts an attention whereby tourism becomes a catalyst for subsequent change. As noted earlier in the book, if Butler's (1980) destination life cycle can apply to resorts in general, so too it might apply to specific red light districts. Moreover, in the communications age, the commodification of the body through cyber-sex tourism means that the potential for the state to effectively control sex is becoming even more doubtful. Ironically, the various measures proposed by governments to control sex sites serves to commodify bodies even more. For example, despite the existence of software content such as Net Nanny which can restrict access by children and the detailed agreements regarding age and intention on many Internet pornography sites, governments seek to restrict access further through utilising age checks such as cyber identification procedures or credit cards – both of which increase the cost of access and the commercialisation of sexuality on the web. Arguably, such commoditisation of sex on the Internet may only further reinforce the creation of a global economy of pleasure of which sex tourism is a part.

Alternative ways of seeing non-commoditised inter-personal relationships are possible but they lie outside of the domains of the dominant culture of contemporary global capitalism. Establishing sites of opposition to the commodifying culture of capitalism is important. The dialectical relationships which provide the means to influence directions, create tensions, and therefore generate new spaces within which relationships can be founded. The liminoid nature of sex tourism, which is by its nature a conceptualisation of shifting boundaries, unclear definitions, differing constructions and disjunctions in society, also provides the possibility for developing new relationships and understandings. Within the global tourism economy of pleasure sex workers need to be recognised as professional providers of sex services in order to ensure that they are not exploited. Such a recognition implies that global measures such as trading and labour agreements therefore provide a significant base with which to improve the conditions of sex workers. However, it also implies that prostitution must be recognised as work.

Discussion and investigation of sex tourism and the wider links between sexuality and tourism in the commodification of the body may also become a catalyst for subsequent change:

> In a period of academic discourse that celebrates 'posts' (postmodernism, postcolonialism, postnationalism) and a new sense of hybrid positionalities and fluidity of categories, it is imperative that we give adequate weight to the intransigence of physical evidence in systems of social differentiation and track these operations in public discourse. Tracking that trace is the only way potentially to disrupt it and to reconfigure the possible meanings of bodily presence.
>
> (Desmond, 1999: 266)

'Recognizing sexual alienation as part of a totalizing system makes it hard to accept easy answers about what is to be done about sex tourism, because all are, at best, partial solutions' (Bishop and Robinson, 1998: 249). Profound change can only occur within a wholesale transformation of economic, social, gender, and political relations. The issues of sex tourism need to be seen as part of wider questions of sexuality, commoditisation, and tourism and the politics of moral and sexual control. However, such transformations require a starting point which, at the very least, requires recognition that sex tourism is part of a series of relationships between supply and demand, worker and client, the individual and society, and that the rights of workers can no longer be ignored.

8 Afterword

From the perspective of a non-positivistic research tradition, any reporting of research is eloquent when the text is re-construed from a stance of examining both that which is said, and that which remains silent. In that sense the writing of a book about matters as controversial as sex tourism is in itself a difficult task. As Manderson (1992) noted, it leaves the authors exposed, even while judgements on the authors' work convey meaning about those who make comment. Further, to include in a book research undertaken at a primary level that was based on, in some cases, comparatively long relationships with informants, makes the task all the more difficult. As has been discussed, the authors have sought to examine on the one hand, issues of motive and, on the other, their relationship with informants. In the previous chapter, one value judgement that is a result of this work has been made explicit, and that is that women possess a right to work as sex workers if they so wish, albeit it is recognised that the conditions that give rise to 'choice' may be constrained by economic necessities among other factors. In short, the authors deny that all sex workers are, by definition, victims. Yet, as hopefully is made clear by the chapter on sex trafficking, it is equally recognised that power structures exist whereby there is undoubtedly exploitation. Legalisation and/or decriminalisation (as may be deemed appropriate by those involved) of the industry is not seen as an answer to all the problems, but it is seen as a means of securing better support mechanisms in terms of health, protection and self-assessment by those men and women who work as sex workers. Such a course of action runs the danger of perpetuating those systems that create the conditions for sex tourism, but from a purely pragmatic perspective the authors would ask, how many women (and men) need to continue in danger from middle men who withhold monies, health protection and can continue to threaten sex workers, while the wider social problems of inequitable distribution of economic power are 'put right'?

In the act of writing though, what has thus far been left unsaid is the relationship between writer and reader. If the authors have had to deal with considerations of voyeurism, where does that leave the reader? Sex is about intimate matters of self-identity as explored in Chapter 4, and therefore can any reader come to an issue such as sex tourism with a mind stripped clean of perceptions, judgements and a sense of their own relationship to the topic, whether it be curiosity, fascination or seeking to clarify their own views? Through adopting a conceptualisation of margins, the authors have sought to develop a series of tensions that portray sex work and sex tourism as paradoxically simultaneously embedded in and yet marginalised by wider society. Margins, it was argued, illustrate by drawing boundaries and illuminating alternatives. They are additionally multi-layered in meaning, and it is hoped that this book has shown that sex tourism as it is known today has historical antecedents in the nineteenth century that have cast long shadows. It was not the intention of the authors when they commenced this project, to engage in a polemic. Hopefully the reader feels that this has not been the case. But just as the authors have felt compelled to eventually establish a judgement, so too, it is hoped, the reader will be better able to make their own judgements, be that what they will, as a consequence of this text.

Bibliography

ABC News (1999) 'Mardi Gras opening draws tens of thousands', *ABC News*, 6 February.

ABC Newslink (1999) 'Gay lobby rejects anti-gay tourism comments', ABC Newslink, 1 February, (7:09 a.m. AEDT), http://www.abc.net.au/news/newslink/daily/newsnat-feb1999–13.htm

Abramson, A. (1993) 'Between autobiography and method: being males, seeing myth and the analysis of structures of gender and sexuality in the eastern interior of Fiji', in D. Bell, P. Caplan and W.J. Karim (eds) *Gendered Fields: Women, Men and Ethnography*, London: Routledge, pp. 63–77.

Albuquerque, de, K. (1998) 'Sex, beach boys and female tourists in the Caribbean', *Sexuality and Culture* 2: 87–111.

—— (1999) 'In search of the big bamboo', *Transition* 77 (March): 48–57.

Anderson, C. (1997) *Our Man in ...*, London: BBC Books.

Ansley, B. (2000) 'Sex and the city', *The New Zealand Listener* 1–7 April: 18–23.

Anti-Slavery International (1995) *Redefining Prostitution as Commercial Sex Work on the International Agenda*, London: Anti-Slavery International.

Ap, J. (1992) 'Residents' perceptions of tourism impacts', *Annals of Tourism Research* 19(4): 665–690.

Aramaberri, J. (2000) 'Night market', *Annals of Tourism Research* 27(1): 241–243.

Ashworth, G.J., White, P.E. and Winchester, H. (1988) 'The redlight district of the West European city: a neglected aspect of the urban landscape', *Geoforum* 19: 201–212.

Assavanonda, A. (1996) 'Suppression officials briefed on new bill to fight flesh trade – main aim is not to punish prostitutes', *Bangkok Post* 21 December, http://www.bkkpost.samart.co.th/news/Bparchive/BP961221/2112_news07.htm (accessed 21 December).

Austin, M. (1998) *Dancing Naked: An Exhibitionist Revealed*, Auckland: Random House.

Bacani, C. (1998) 'The perfect vacation', *AsiaWeek*, 6 November.

Baillie, J.G. (1980) 'Recent international travel trends in Canada', *Canadian Geographer* 24(1): 13–21.

Baring, A. and Cashford, J. (1991) *The Myth of the Goddess: Evolution of an Image*, London: BCA.

Barry, K. (1984) *Female Sexual Slavery*, New York: New York University Press.

—— (1995) *The Prostitution of Sexuality*, New York: New York University Press.

Bauman, Z. (1994) 'Fran pilgrim till turist (From Pilgrim to Tourist)', *Moderna Tider* September: 20–34.

Beddoe, C. (1998) 'Beachboys and tourists: links in the chain of child prostitution in Sri Lanka', in M. Oppermann (ed.) *Sex Tourism and Prostitution: Aspects of Leisure, Recreation, and Work*, New York: Cognizant Communication Corporation, pp. 45–59.

Begg, Z. (1999) 'Lesbian and gay rights: is Mardi Gras enough?', *Green Left Weekly*, 24 February.

Bell, S. (1994) *Reading, Writing and Rewriting the Prostitute Body*, Bloomington and Indianapolis: Indiana University Press.

Bem, S. (1993) *The Lenses of Gender: Transforming the Debate on Sexual Inequality*, New Haven, CT: Yale University Press.

Benny Dorm (pseudonym) (1988) *Beach Party – The Last Resort*, London: Hodder and Stoughton.

The Best Guide to Amsterdam and Benelux (1992) fourth edition, Amsterdam: Bookscene.

Bishop, R. and Robinson, L.S. (1998) *Night Market: Sexual Cultures and the Thai Economic Miracle*, London and New York: Routledge.

Blanch, L. (1984) *The Wilder Shores of Love*, London: Abacus.

Boorstin, D.J. (1963) *The Image, or What Happened to the American Dream?*, Harmondsworth: Penguin Books.

Boskin, M.J. (1996) 'National satellite accounting for travel and tourism: a cold review of the WTTC/WEFA group research', *Tourism Economics* 2(1): 3–12.

Bowman, G. (1988) 'Impacts of tourism', paper presented at the Conference on the Anthropology of Tourism, Froebel College, London.

Boyle, S. (1994) *Working Girls and their Men*, London: Smith Gryphon Publishers.

Brierley, H. (1984) 'Gender identity and sexual behaviour', in K. Howells (ed.) *The Psychology of Sexual Diversity*, Oxford: Basil Blackwell, pp. 63–88.

Broadbent, L. (1998) 'Open wide, come inside, it's love school', *Cleo* August: 36–41.

Brother Sister (1995) 'Tourism industry backs conference', *Brother Sister – Queer News From Downunder* 79, 4 May.

—— (1996a) 'Pink dollar hot again', *Brother Sister – Queer News From Downunder* 122, 26 December.

—— (1996b) 'Tas laws cost tourism', *Brother Sister – Queer News From Downunder* 119, 14 November.

—— (1996c) 'Shameful human rights record', *Brother Sister – Queer News From Downunder* 121, 12 December.

Bucknell, J. (1999) 'Review', *The Weekend Australian*, 6 February.

Burley, N. and Symanski, R. (1981) 'Women without: an evolutionary and cross-cultural perspective on prostitution', in R. Symanski (ed.) *The Immoral Landscape: Female Prostitution in Western Societies*, Toronto: Butterworths, pp. 273–293.

Burrell, I. (1997) 'Happy hookers declare war on feminist writers', *The Independent International* 17–23 December: 20.

Butler, R.W. (1980) 'The concept of a tourism area cycle of evolution', *Canadian Geographer* 24(1): 5–12.

Cabezas, A.L. (1999) 'Women's work is never done: sex tourism in Sosua, the Dominican Republic', in K. Kempadoo (ed.) *Sun, Sex, and Gold – Tourism and Sex Work in the Caribbean*, Lanham, MD: Rowman and Littlefield Publishers Inc., pp. 93–124.

Cameron, A. (1988) *Daughters of Copper Woman*, Press Gang Publishers.

Canadian Press (1999) 'Gays prefer marriage over single life: poll', *Canadian Press*, Montreal, 11 February.

Caplan, G.M. (1984) 'The facts of life about teenage prostitution', *Crime and Delinquency* 30(1): 69–74.

Carlyle, T. (1965/1843) *Past and Present*, London: Oxford University Press.

Chapkis, W. (1997) *Live Sex Acts: Women Performing Erotic Labour*, New York: Routledge.

Claire, R. and Cottingham, J. (1982) 'Migration and tourism: an overview', in *Women in Development: A Resource Guide for Organisation and Action*, Geneva: ISIS Women's International and Communication Service, pp. 205–215.

Clark, N. and Clift, S. (1996) 'Dimensions of holiday experiences and their health implications: a study of British tourists in Malta', in S. Clift and S. Page (eds) *Health and the International Tourist*, London: Routledge, pp. 108–133.

Clift, S. and Forrest, S. (1999) 'Gay men and tourism: destinations and holiday motivations', *Tourism Management* 20: 615–625.

Clift, S. and Page, S.J. (1996) *Health and the International Tourist*, London: Routledge.

Cockburn, R. (1988a) 'The geography of prostitution, part I: the east', *The Geographical Magazine* 60(3): 2–5.

—— (1988b) 'The geography of prostitution, part II: the west', *The Geographical Magazine* 60(4): 44–47.

Cohen, E. (1971) 'Arab boys and tourist girls in a mixed Jewish-Arab community', *International Journal of Comparative Sociology* 12: 217–233.

—— (1979) 'Rethinking the sociology of tourism', *Annals of Tourism Research* 6(1): 18–35.

—— (1982) 'Thai girls and farang men: the edge of ambiguity', *Annals of Tourism Research* 9: 403–428.

—— (1986) 'Lovelorn farangs: the correspondence between foreign men and Thai girls', *Anthropological Quarterly* 59(3): 115–127.

—— (1988) 'Tourism and AIDS in Thailand', *Annals of Tourism Research* 15: 467–486.

—— (1993) 'Open-ended prostitution as a skilful game of luck: opportunity, risk and security among tourist-oriented prostitutes in a Bangkok soi', in M. Hitchcock, V.T. King and M.J.G. Parnwell (eds) *Tourism in South-East Asia*, London: Routledge, pp. 155–178.

Cohen, L.E. and Felson, M. (1979) 'Social change and crime rate trends: a routine activity approach', *American Sociological Review* 44: 588–608.

Commonwealth of Australia Gazette (1994) 'Legislation', *Commonwealth of Australia Gazette*, No. GN28, 20 July: 1479.

Community Marketing (1999) *Annual Gay and Lesbian Travel Survey*, San Francisco: Community Marketing.

Cooper, M. and Hanson, J. (1998) 'Where there are no tourists … yet: a visit to the slum brothels in Ho Chi Minh City, Vietnam', in M. Oppermann (ed.) *Sex Tourism and Prostitution: Aspects of Leisure, Recreation and Work*, New York: Cognizant Communication Corporation, pp. 144–152.

Cornish, D.B. and Clarke, R.V. (1986) *The Reasoning Criminal – Rational Choice Perspectives on Offending*, New York: Springer Verlag.

Craik, J. (1997) 'The culture of tourism', in C. Rojek and J. Urry (eds) *Touring Cultures – Transformations of Travel and Theory*, London: Routledge, pp. 113–136.

Crompton, J.L. (1979) 'Motivation for pleasure vacation', *Annals of Tourism Research* 6(4): 408–424.

Cross, G. (1989) 'Vacations for all: the leisure questions in the era of the popular front', *Journal of Contemporary History* 24: 599–621.

Currie, R.R. (1997) 'A pleasure-tourism behaviors framework', *Annals of Tourism Research* 24: 884–897.

Daniel, N. (1996) *Vietnamese children sold into sex slavery*, http://www.vinsight.org/1996news/1204a.htm

Daniels, C. (1999) 'Anti-Hero Parade ad stirs protest', *The New Zealand Herald*, 12 February.

Davidson, D. (1985) 'Women in Thailand', *Canadian Women's Studies* 16(1): 16–19.

Delacoste, F.D. and Alexander, P. (1987) *Sex Work, Writings by Women in the Sex Industry*, Pittsburgh: Cleis Press.

Desmond, J.C. (1999) *Staging Tourism: Bodies on Display From Waikiki to Sea World*, Chicago: University of Chicago Press.

Douglas, N. (1996) *They Came for Savages: 100 Years of Tourism in Melanesia*, Lismore: Southern Cross University Press.

Douglas, N. and Douglas, N. (1996) 'Tourism in the Pacific: historical factors', in C.M. Hall and S. Page (eds) *Tourism in the Pacific: Issues and Cases*, London: International Thomson Business Press, pp. 19–35.

Downes, D. and Rock, P. (1988) *Understanding Deviance – a Guide to the Sociology of Crime and Rule-Breaking*, 2nd edition, Oxford: Clarendon Press.

Dragu, M. and Harrison, A.S.A., 1989, *Revelations: Essays on Striptease and Sexuality*, London (Ontario): Nightwood Editions.

Durkheim, E. ([1893] 1933) *The Division of Labour in Society*, Chicago: Free Press.

Dworkin, A. (1988) *Letters from a War Zone: Writings 1976–1987*, London: Martin Secker and Warburg.

Easson, Mrs (1994) *House of Representatives, Hansard* 4 May: 146.

ECPAT (nd) *What is child sex tourism?*, http://ecpat/org (accessed January 1999).

ECPAT Australia (1996a) 'Update: arrests and convictions', *ECPAT Australia* 35 (October): 6.

—— (1996b) 'Legal updates', *ECPAT Australia* 28 (February): 1.

—— (1996c) 'Minister of Foreign Affairs orders a "ruthless" inquiry into allegations of child sex abuse and Australian diplomatic staff', *ECPAT Australia* 30 (April): 1.

—— (1996d) 'The N.S.W. Royal Commission continues to shock Australia and rock institutions', *ECPAT Australia* 30 (April): 3.

—— (2000) 'Australia's National Plan of Action', *ECPAT Australia* 55 (February): 1.

Edensor, T. and Kothari, U. (1994) 'The masculinisation of Stirling's heritage', in V.H. Kinnaird and D.R. Hall (eds) *Tourism: A Gender Analysis*, Chichester: John Wiley, pp. 164–187.

Edwards, A. (1997) 'Vietnam says tourism leading to crime', Reuters, Hanoi, 24 September.

Edwards, S. (1997) 'The legal regulation of prostitution: a human rights issue', in G. Scambler and A. Scambler (eds) *Rethinking Prostitution: Purchasing Sex in the 1990s*, London: Routledge, pp. 57–82.

English Collective of Prostitutes (1997) 'Campaigning for legal change', in G. Scambler and A. Scambler (eds) *Rethinking Prostitution: Purchasing Sex in the 1990s*, London: Routledge, pp. 83–104.

Enloe, C. (1990) *Bananas, Beaches and Bases: Making Feminist Sense of International Politics*, Berkeley, CA: University of California Press.

—— (1992) 'It takes two', in S.P. Sturdevant and B. Stolzfus (eds) *Let the Good Times Roll: Prostitution and the U.S. Military in Asia*, New York: The New Press, pp. 22–27.

Ericsson, L. (1980) 'Charges against prostitution: an attempt at a philosophical assessment', *Ethics* 90: 335–366.

Express (1999) 'What is HERO?', *Express*, 4 February.

Faugier, J. (1994) 'Bad women and good customers, scapegoating, female prostitution and HIV', in C. Webb (ed.) *Living Sexuality: Issues for Nursing and Health*, London: Scutari Press, pp. 50–64.

Faugier, J. and Sargeant M. (1997) 'Boyfriends, pimps, and clients', in G. Scambler and A. Scambler (eds) *Rethinking Prostitution: Purchasing Sex in the 1990s*, London: Routledge, pp. 121–138.

Felson, M. (1986) 'Linking criminal choices, routine activities, informal control, and criminal outcomes', in D.B. Cornish and R.V. Clarke (eds) *The Reasoning Criminal – Rational Choice Perspectives on Offending*, New York: Springer-Verlag.

Fish, M. (1984a) 'Deterring sex sales to international tourists: A case study of Thailand, South Korea and the Philippines', *International Journal of Comparative and Applied Criminal Justice* 8(2): 175–186.

Fish, M.M. (1984b) 'On controlling sex sales to tourists: commentary on Graburn and Cohen', *Annals of Tourism Research* 11(4): 615–617.

Fletcher, S.D. (1996) *e-mail//a.love.story//*, London: Headline Book Publishing.

Ford, N. and Eiser, J.R. (1996) 'Risk and liminality: the HIV-related socio-sexual interaction of young tourists', in S. Clift and S. Page (eds) *Health and the International Tourist*, London: Routledge, pp. 152–178.

Foucault, M. ([1976] 1990), *The History of Sexuality. Volume 1: An Introduction*, London: Penguin Books.

—— ([1977] 1996) 'The Carceral', reprinted in J. Farganis (ed.) *Readings in Social Theory: The Classic Tradition to Post-Modernism*, New York: McGraw-Hill.

Garfinkel, H. (1967) *Studies in Ethnomethodology*, Englewood Cliffs, NJ: Prentice-Hall.

Gay, J. (1985) 'The patriotic prostitute', *The Progressive* 49(3): 34–36.

Gibson-Ainyette, I., Templer, D.I., Brown, R. and Veaco, L. (1988) 'Adolescent female prostitutes', *Archives of Sexual Behaviour* 17(5): 431–438.

Gillies, P. and Slack, R. (1996) 'Context and culture in HIV prevention: the importance of holidays', in S. Clift and S. Page (eds) *Health and the International Tourist*, London: Routledge, pp. 134–151.

Goffman, E. (1967) *Interaction Ritual: Essays on Face to Face Behaviour*, Garden City, NY: Anchor Books.

Goldman, E. ([1917] 1970) *The Traffic in Women and Other Essays on Feminism*, Ojai, CA: Times Change Press.

Goldsmith, M. (1996) *Political Incorrectness – Defying the Thought Police*, Rydalemere, NSW: Hodder and Stoughton.

Graburn, N.H.H. (1983a) 'The anthropology of tourism', *Annals of Tourism Research* 10: 9–33.

—— (1983b) 'Tourism and prostitution', *Annals of Tourism Research* 10: 437–456.

Gray, S. (*c.* 1985) *Swimming to Cambodia*, New York: Theatre Communications Group.

Grayman, J., Williams, A., Young, P.E. and Yeager, A.E. (1996) *Let's Go: The Budget Guide to Southeast Asia*, New York: St Martin's Press.

Grey, T. C. (1983) *The Legal Reinforcement of Morality*, New York: Knopf.

Guba, E.G. (ed.) (1990) *The Paradigm Dialogue*, Newbury Park: Sage Publications.

Günther, A. (1998) 'Sex tourism without sex tourists', in M. Oppermann (ed.) *Sex Tourism and Prostitution: Aspects of Leisure, Recreation and Work*, New York: Cognizant Communication Corporation, pp. 71–80.

Haine, W.S. (1992) 'The development of leisure and the transformation of working class adolescence, Paris 1830–1940', *Journal of Family History* 17(4): 451–476.

Hall, C.M. (1992) 'Sex tourism in South-East Asia', in D. Harrison (ed.) *Tourism and the Less Developed Nations*, London: Belhaven Press, pp. 64–74.

—— (1994) 'Nature and implications of sex tourism in South-East Asia', in V.H. Kinnaird and D.R. Hall (eds) *Tourism: A Gender Analysis*, Chichester: John Wiley, pp. 142–163.

—— (1996) 'Tourism prostitution: the control and health implications of sex tourism in South-East Asia and Australia', in S. Clift and S. Page (eds) *Health and the International Tourist*, London: Routledge, pp. 179–197.

—— (1998) 'The legal and political dimensions of sex tourism: the case of Australia's child sex tourism legislation', in M. Oppermann (ed.) *Sex Tourism and Prostitution: Aspects of Leisure, Recreation, and Work*, New York: Cognizant Communication Corporation, pp. 87–96.

—— (1999) 'Voices from the periphery: recent Australasian writings on tourism and related subjects', *Tourism Geographies* 1(2): 244–250.

Hall, C.M. and Page, S. (eds) (2000) *Tourism in South and South East Asia*, Oxford: Butterworth-Heinemann.

Hamilton, A. (1997) 'Primal dream: masculinism, sin and salvation in Thailand's sex trade', in L. Manderson and M. Jolly (eds) *Sites of Desire, Economies of Pleasure: Sexualities in Asia and the Pacific*, Chicago: University of Chicago Press, pp. 145–165.

Hampden, T.C. (1971) *Radical Man*, London: Duckworth.

Hampson, S. (1986) 'Sex roles and personality', in D.J. Hargreaves and A.M. Colley (eds) *The Psychology of Sex Roles*, London: Harper and Row, pp. 45–59.

Hanna, J.L. (1998) 'Undressing the first amendment and corsetting the striptease dancer', *The Dram Review – A Journal of Performance Studies* 42(2): 38–69.

Hanson, J. (1994–5) *Notes from New Zealand*, thirty reports for the Canadian Broadcasting Corporation's (Saskatchewan) Morning Edition.

—— (1996a) 'Sex tourism in New Zealand: what do the Kiwi prostitutes have to say about it?', in M. Oppermann (ed.) *Pacific Rim Tourism 2000, Issues, Interrelations, Inhibitors – Conference Proceedings*, Rotorua, New Zealand: Centre for Tourism Studies, Waiariki Polytechnic, pp.105–113.

—— (1996b) 'Learning to be a prostitute: education and training in the New Zealand Sex Industry', paper prepared for the *Women's Studies Journal*, Department of Education Studies, University of Waikato.

—— (1998) 'Child prostitution in South East Asia: white slavery revisited?', in M. Oppermann (ed.) *Sex Tourism and Prostitution: Aspects of Leisure, Recreation and Work*, New York: Cognizant Communication Corporation, pp. 51–59.

Harrell-Bond, B. (1978) *'A window on an outside world': Tourism as Development in the Gambia*, American Universities Field Staff Reports No. 19, Hanover: American Universities Field Staff.

Hart, A. (1995) '(Re)constructing a Spanish red-light district; prostitution, space and power', in D. Bell and G. Valentine (eds) *Mapping Desire*, London: Routledge, pp. 214–230.

Hawkesworth, M. (1984) 'Brothels and betrayal: on the functions of prostitution', *International Journal of Women's Studies* 7(1): 81–91.

Hill, C. (1993) 'Planning for prostitution: an analysis of Thailand's sex industry', in M. Turshen and B. Halcomb (eds) *Women's Lives and Public Policy: The International Experience*, Westport, Greenwood Press, pp.133–144.

Hirshi, T. (1962) 'The professional prostitute', *Berkeley Journal of Sociology* 7: 33–49.

Hobson, J.P.S. and Heung, V. (1998) 'Business travel and the emergence of the modern Chinese concubine', in M. Oppermann (ed.) *Sex Tourism and Prostitution: Aspects of Leisure, Recreation and Work*, New York: Cognizant Communication Corporation, pp. 132–143.

Høigård, C. and Finstad, L. (1992). *Backstreets: Prostitution, Money and Love*, Cambridge: Polity Press.

Holcomb, B. and Luongo, M. (1996) 'Gay tourism in the United States', *Annals of Tourism Research* 23(3): 711–713.

Hollinshead, K. (1997) 'Heritage tourism under post-modernity: truth and the past', in C. Ryan (ed.) *The Tourist Experience: A New Introduction*, London: Cassell.

—— (1999) 'Surveillance of the worlds of tourism: Foucault and the eye of power', *Tourism Management* 20(1): 7–24.

Hopkins, R. (1995) 'Mardi Gras gets tourism praise', *Brother Sister – Queer News From Downunder* 94, 30 November.

Hornery, A. (1999) 'Blue chips jump onto the pink wagon', *Sydney Morning Herald*, 22 February.

Hosein, E. (1995) 'C'bean men sex jokers', *The Barbados Advocate*, 21 February :15.

House of Representatives (1994) *Hansard*, 29 June: 2345.

House of Representatives Standing Committee on Legal and Constitutional Affairs [HRSCLCA] (1994) *Crimes (Child Sex Tourism) Amendment Bill 1994 Advisory Report*, Canberra: House of Representatives Standing Committee on Legal and Constitutional Affairs, Parliament of the Commonwealth of Australia.

Howard, J.A. and Hollander, J.A. (1997) *Gendered Situations, Gendered Selves*, Thousand Oaks, CA: Sage Publications.

Hoy, D.C. (1986) *Foucault: A Critical Reader*, Oxford: Blackwell.

Huckshorn, K. (1996) 'Vietnamese cops raid foreigners' bar – incident highlights friction between West, Hanoi government', *San Jose Mercury News*, 30 May.

Hughes, D. (1999) 'The Internet and the global prostitution industry', in S. Hawthorne and R. Klein (eds) *The Internet and the Global Prostitution Industry*, North Melbourne: Spinifex, pp. 157–184.

Hughes, H. (1997) 'Holidays and homosexual identity', *Tourism Management* 18(1): 3–9.

Human Rights Watch (1999) *Trafficking*, http://www.hrw.org/about/projects/traffcamp/intro.html (accessed 25 January, 2000).

International Labour Organisation (ILO) (1995) 'ILO expert declares "widespread" forced child prostitution triggers trauma', *ILO Washington Focus* Fall.

IRN News (1999) 'Christians want boycott of retailers linked to Hero parade', *IRN News*, New Zealand, 12 February.

ISIS (1979) *Tourism and Prostitution*, International Bulletin 13, Geneva: ISIS.

Jackson, P.A. (1997) 'Kathoey Gay Man: the historical emergence of gay male identity in Thailand', in L. Manderson and M. Jolly (eds) *Sites of Desire, Economies of Pleasure: Sexualities in Asia and the Pacific*, Chicago: University of Chicago Press, pp. 166–190.

J-FLAG (1999) 'Concerns re Jamaica's homophobia and tourist market', *Press Release*, J-FLAG, 10 February, http://village.fortunecity.com/garland/704/ (accessed 25 January, 2000).
James, W. (1890/1991) *The Principles of Psychology*, Vol. 1, Cambridge, MA: Harvard University Press.
Jaschok, M. and Miers, S. (1994) 'Traditionalism, continuity and change' in M. Jaschok and S. Miers (eds), *Women and Chinese Patriarchy: Submission, Servitude and Escape*, London: Routledge, pp 264–267.
Jeffreys, S. (1997) *The Idea of Prostitution*, North Melbourne: Spinifex Press.
Jones, M. (1986) *A Shady Place for Shady People: The Real Gold Coast Story*, Sydney: Allen and Unwin.
Jordan, J. (1991) *Working Girls: Women in the New Zealand Sex Industry Talk to Jan Jordan*, Auckland: Penguin.
Kanato, M. and Rujkorankarn, D. (1994) 'Cultural factors in sexual behaviour, sexuality and sociocultural contexts in the spread of HIV in the Northeast Thailand', paper presented at workshop, Cultural Dimensions of AIDS Control and Care in Thailand, Chiangmai.
Kasl, C.D. (1989) *Women, Sex and Addiction*, New York: Ticknor and Fields.
Kempadoo, K. (1999) 'Continuities and change: five centuries of prostitution in the Caribbean', in K. Kempadoo (ed.) *Sun, Sex, and Gold: Tourism and Sex Work in the Caribbean*, Lanham, MD: Rowman and Littlefield, pp. 3–36.
Kempadoo, K. and Doezema, J. (eds) (1998) *Global Sex Workers: Rights, Resistance and Redefinition*, London: Routledge.
Kempadoo, K. and Ghuma, R. (1999) 'For the children: trends in international policies and law on sex tourism', in K. Kempadoo (ed.) *Sun, Sex, and Gold: Tourism and Sex Work in the Caribbean*, Lanham. MD: Rowman and Littlefield, pp. 291–308.
Kerr, D. (1994) *Crimes (Child Sex Tourism) Amendment Bill 1994, Second Reading Speech*, Canberra: House of Representatives, Parliament of the Commonwealth of Australia.
Khazar, M. (1998) 'State of decay', *Penthouse* (New Zealand Issue) 19(10) October.
King, K. (1994) *Theory in Its Feminist Travels: Conversations in U.S. Women's Movements*, Indiana: Indiana University Press.
Kinnaird, V.H. and Hall, D.R. (eds) (1994) *Tourism: A Gender Analysis*, Chichester: John Wiley.
Kleiber, D. and Wilke, M. (1995) *Prostitutionskunden: Eine Untersuchung über soziale und psychologische Charakteristike von Besuchern weiblicher Prostituierten in Zeiten von AIDS*, Baden-Baden: Nomos-Verlagsgesellschaft.
Klerk, Y. (1995) 'Definition of "traffic in persons"', in M. Klap, Y. Klerk and J. Smith (eds) *Combatting Traffic in Persons*, SIM Special no.17, Utrecht: Studie – en Informatiecentrum Mensenrechten.
Kohm, S. and Selwood, J. (1998) 'The virtual tourist and sex in cyberspace', in M. Oppermann (ed.) *Sex Tourism and Prostitution: Aspects of Leisure, Recreation and Work*, New York: Cognizant Communication Corporation, pp. 123–131.
Kristof, N.D. (1996) 'Children victimized by Asian brothels', 16 April, http://www.vinsight.org/1996news/1204a.htm
Kruhse-MountBurton, S. (1995) 'Sex tourism and traditional Australian male identity', in M-F. Lanfant, J.B. Allcock and E.M. Bruner (eds) *International Tourism – Identity and Change*, London: Sage.

—— (1996) 'The contemporary client of prostitution in Darwin, Australia', unpublished PhD thesis, Griffith University, Nathan, Queensland.

Laga, T. (1993) 'Pelacuran versus moralisasi' (Prostitution versus morality), *Pos Kupang*, 22 March, Nusa Tenggara, Timur, Indonesia.

Lamont-Brown, R. (1982) 'The international expansion of Japan's criminal brotherhood', *Police Journal* 55(4): 355–359.

Lap-Chew, L. (1995) 'Letter to the special rapporteur', *Foundation Against Trafficking in Women: News Bulletin* 2(March).

Lawrence, D. (1982) 'Parades, politics, and competing urban images: doo dah and roses', *Urban Anthropology* 11: 155–176.

Lea, J. (1993) 'Tourism development ethics in the Third World', *Annals of Tourism Research* 20: 701–715.

Lee, J. (1988) 'Care to join me in an upwardly mobile tango?: postmodernism and the "new woman"', in L. Gamman and M. Marshment (eds) *The Female Gaze*, London: The Women's Press.

Leheny, D. (1995) 'A political economy of Asian sex tourism', *Annals of Tourism Research* 22: 367–384.

Leonowens, A.H. ([1873] 1953) *Siamese Harem Life*, New York: E.P. Dutton and Co.

—— ([1873] 1991) *The Romance of the Harem*, ed. S. Morgan. Charlottesville: University Press of Virginia.

Lett, J.W. (1983) 'Ludic and liminoid aspects of charter yacht tourism in the British Virgin Islands', *Annals of Tourism Research* 10: 35–56.

Little, J. (1999) 'Sex tourists and sex workers in the Caribbean', unfinished PhD thesis, School of Business, University of Luton.

Lodge, D. (1992) *Paradise News*, London: Penguin

—— (1995) *Therapy – A Novel*, London: Secker and Warburg.

Lofland, J. (1971) *Analysing Social Situations*, Belmont: Wadsworth.

MacCannell, D. (1976) *The Tourist: A New Theory of the Leisure Class*, New York: Schocken.

—— (1992) *Empty Meeting Grounds: The Tourist Papers*, London: Routledge.

McDonald-Leigh, S. (1999) 'What Mardi Gras means to me', *Sydney Citysearch*, 5 February, http://203.26.177.133/E/F/SYDNE/0000/05/00/ (accessed 25 January, 2000).

McGuire, R.J., Carlisle, J.M. and Young, B.G. (1965) 'Sexual deviation as conditioned behaviour', *Behaviour Research and Therapy* 2: 185–190.

McLachlan, D. (1999) 'The meaning of Mardi Gras', *Sydney Star Observer*, 4 February.

McLeod, E. (1982) *Women Working: Prostitution Now*, London, Croom Helm.

Mager, D. and Andrews J. (2000) 'Trade in illegal migrants smashed: debtors forced to crime', *The New Zealand Herald*, 14 March: 1.

Malinowski, B. (1922) *Argonauts of the Western Pacific*, London: Routledge and Kegan Paul.

Manderson, L. (1992) 'Public sex performances in Patpong and explorations of the edges of imagination', *The Journal of Sex Research* 29(4): 451–475.

—— (1997) 'Parables of imperialism and fantasies of the exotic: Western representations of Thailand – place and sex', in L. Manderson and M. Jolly (eds) *Sites of Desire, Economies of Pleasure: Sexualities in Asia and the Pacific*, Chicago: University of Chicago Press, pp. 123–144.

Marcovich, M. (1998) 'Prostitution is violence against humankind', speech given in Norway, June, http://www.uri.edu/artsci/wms/hughes/catw/marcono.htm (accessed September,1998).

Markwell, K. (1998) 'Taming the "chaos of nature": cultural construction and lived experience in nature-based tourism', unpublished PhD thesis, University of Newcastle, New South Wales.

Marsh, C. (1997) 'The spirituality of Shirley Valentine', in C. Marsh and G. Ortiz (eds) *Explorations in Theology and Film: Movies and Meaning*, Oxford: Blackwell.

Marshment, M. (1997) 'Gender takes a holiday: representation in holiday brochures', in T. Sinclair (ed.) *Gender, Work and Tourism*, London: Routledge, pp. 16–34.

Marx, K. ([1844] 1964) *Economic and Philosophic Manuscripts of 1844*, New York: International Publishers.

Marx, K. and Engels, F. ([1844] 1993) 'The manifesto of the Communist Party', in J. Farganis (ed.) *Readings in Social Theory – The Classic Tradition to Post-modernism*, New York: McGraw-Hill.

Maslow, A. (1970) *Motivation and Personality*, 2nd edition, New York: Harper and Row.

Mathieson, A. and Wall. G. (1982) *Tourism: Economic, Social, and Environmental Impacts*, Harlow: Longmans.

Matsui, Y. (1987a) 'The prostitution areas in Asia: an experience', *Women in a Changing World* 24: 27–32.

—— (1987b) 'Japan in the context of the militarisation of Asia', *Women in a Changing World* 24: 7–8.

—— (1999) *Women in the New Asia*, London: Zed Books.

Matthews, H.G. (1978) *International Tourism: A Political and Social Analysis*, Cambridge: Schenkman Publishing Company.

Matthews, R. (1997) *Prostitution in London: An Audit*, monograph, London: Department of Social Science, Middlesex University.

Mayhew, H. ([1851] 1999) *London Labour and the London Poor*, London: Folio Society.

Mead, M. (1977) *Letters from the Field 1925–1975*, New York: Harper and Row.

Miles, R. (1992) *The Rites of Man: Love, Sex and Death in the Making of the Male*, London: Paladin.

Miller, R. (1998) 'Rejoinder – from the World Travel and Tourism Council', *Tourism Economics* 4(1): 71–78.

Milner, C. and Milner, R. (1972) *Black Players*, Boston: Little and Brown.

Momocco, M. (1998) 'Japanese sex workers: encourage, empower, trust and love yourselves', in K. Kempadoo and J. Doezema (eds) *Global Sex Workers: Rights, Resistance and Redefinition*, London: Routledge, pp. 178–181.

Montgomery, H. (1998) 'Children, prostitution, and identity: a case study from a tourist resort in Thailand', in K. Kempadoo and J. Doezema (eds) *Global Sex Workers: Rights, Resistance and Redefinition*, London: Routledge, pp. 139–150.

Moore, C.G. (1991) *A Killing Smile*, Bangkok: White Lotus Press.

—— (1993) *A Haunting Smile*, Bangkok: White Lotus Press.

—— (1995) *The Comfort Zone*, Bangkok: White Lotus Press.

—— (1996) *The Big Weird*, Bangkok: Siam Books.

Morgan, S. (1991) 'Introduction', in A. Leonowens, *The Romance of the Harem*, Charlottesville: University Press of Virginia.

Moselina, L. (1979) 'Rest & recreation: the U.S. naval base at Subic Bay', in *Tourism and Prostitution*, International Bulletin 13, Geneva: ISIS, pp. 17–20.

Movement to Prevent Child Prostitution (MPCP) (1996) 'Paedophiles prey on Sri Lankan children', *The Movement to Prevent Child Prostitution (MPCP) News Letter*, Wembley Park: MPCP.

Mullins, B. (1999) 'Globalization, tourism and the international sex trade', in K. Kempadoo (ed.) *Sun, Sex, and Gold – Tourism and Sex Work in the Caribbean*, Lanham, MD: Rowman and Littlefield Publishers Inc., pp. 55–80.

Mulvey, L. (1981) 'Visual pleasure and narrative cinema', in T. Bennett *et al.* (eds) *Popular Film and Television*, London: Open University Press.

Munt, S. (1998) 'The lesbian flaneur', in D. Bell and G. Valentine (eds) *Mapping Desire*, London: Routledge, pp. 114–125.

Muntarbhorn, V. (1993) *Sale of Children: Report Submitted by the Special Rapporteur Appointed in Accordance with the Commission on Human Rights Resolution 1992/76*, E/CN.4/1993/67/Add.1, 9 February.

—— (1996) 'Report from the World Congress Against the Commercial Sexual Exploitation of Children, Sweden, August 27–31, 1996', *ECPAT Australia*, 35 (October): 1–2.

Murray, A. (1998a) 'Femme on the streets, butch in the sheets (a play on whores)', in D. Bell and G. Valentine (eds) *Mapping Desire*, London: Routledge, pp. 66–74.

—— (1998b) 'Debt-bondage and trafficking: don't believe the hype', in K. Kempadoo and J. Doezema (eds) *Global Sex Workers: Rights, Resistance and Redefinition*, London: Routledge, pp. 51–64.

Myo Thet Htoon, Myat Thu, Thein Myint, Min Thwe and Saw Edward Zan (nd) *A Study on the Social and Behavioural Pattern of Commercial Sex Workers Returning from Thailand*, Mimeograph, Yangon: Department of Health.

New Zealand Herald (2000) 'Sweatshop operator gets hefty penalty', 8 April, http://www.nzherald.co.nz/storyqueryprocess.cfm?searchtext=Sivoravong&thesection=&period=year&s=1 (accessed April 1999).

O'Connell Davidson, J. (1995) 'British sex tourists in Thailand', in M. Maynard and J. Purvis (eds) *(Hetero) Sexual Politics*, London: Taylor and Francis.

—— (1998) *Prostitution, Power and Freedom*, London: Polity Press.

—— (2000) 'Sex tourism and child prostitution', in S. Clift and S. Cater (eds) *Tourism and Sex: Culture, Commerce and Coercion*, London: Pinter, pp. 54–73.

O'Connell Davidson, J. and Sanchez Taylor, J. (1999) 'Fantasy Islands: exploring the demand for sex tourism', in K. Kempado (ed.) *Sun, Sex, and Gold – Tourism and Sex Work in the Caribbean*, Lanham, MD: Rowman and Littlefield Publishers Inc., pp. 37–54.

Odzer, C. (1994) *Patpong Sisters – An American Woman's View of the Bangkok Sex World*, New York: Arcade Publishing, Inc.

O'Grady, R. (1981) *Third World Stopover*, Geneva: World Council of Churches.

O'Malley, J. (1988) 'Sex tourism and women's status in Thailand', *Loisir et Société* 11(1): 99–114.

Ong, A. (1985) 'Industrialisation and prostitution in southeast Asia', *Southeast Asia Chronicle* 96: 2–6.

Oppermann, M. (ed.) (1998) *Sex Tourism and Prostitution: Aspects of Leisure, Recreation and Work*, New York: Cognizant Communication Corporation.

Oppermann, M., McKinley, S. and Chon, K-S. (1998) 'Marketing sex and tourism destinations', in M. Oppermann (ed.) *Sex Tourism and Prostitution: Aspects of Leisure, Recreation and Work*, New York: Cognizant Communication Corporation, pp. 20–29.

Parasuraman, A., Zeithaml, V.A. and Berry, L.L (1988) 'SERVQUAL: A multiple-item scale for measuring consumer perceptions of service quality research', *Journal of Retailing* 64 (Spring): 12–37.

Parliamentary Joint Committee on the National Crime Authority (PJCNCA) (1995) *Organised Criminal Paedophile Activity*, Canberra: the Parliamentary Joint Committee on the National Crime Authority, Parliament of the Commonwealth of Australia.

Pateman, C. (1988) *The Sexual Contract*, Cambridge: Polity Press.

Pettafor, E. (1999) 'Mardi Gras kiss of life to the economy', *Australian Financial Review*, 27 February.

Philippine Women's Research Collective (1985) *Filipinas for Sale: An Alternative Philippine Report on Women and Tourism*, Quezon City: Philippine Women's Research Collective.

Phillips, J.L. (1999) 'Tourist-oriented prostitution in Barbados: the case of the beach boy and the white female tourist', in K. Kempadoo (ed.) *Sun, Sex, and Gold – Tourism and Sex Work in the Caribbean*, Lanham, MD: Rowman and Littlefield Publishers Inc., pp. 183–200.

Phillips, J. and Dann, G. (1998) 'Bar girls in central Bangkok: prostitution as entrepreneurship', in M. Oppermann (ed.) *Sex Tourism and Prostitution: Aspects of Leisure, Recreation and Work*, New York: Cognizant Communication Corporation, pp. 60–70.

Phoenix, J. (1995) 'Prostitution: problematizing the definition', in M. Maynard and J. Purvis (eds) *(Hetero) Sexual Politics*, London: Taylor and Francis, pp. 65–77.

Phongpaichit, P. and Baker, C. (1995) *Thailand, Economy and Politics*, Kuala Lumpur: Oxford University Press.

Pizam, A. and Mansfeld, Y. (eds) (1996) *Crime and International Security Issues*, Chichester: Wiley.

Plant, M. (1997) 'Alcohol, drugs and social milieu', in G. Scambler and A. Scambler (eds) *Rethinking Prostitution: Purchasing Sex in the 1990s*, London: Routledge, pp. 164–179.

Plumridge, E., Chetwynd, S.J., Reed, A. and Gifford, S.J. (1997) 'Discourses of emotionality in commercial sex: the missing client voice', *Feminism and Psychology* 7(2): 165–181.

Polly, J. (1999) *The Other Australia*, http://www.gay.net (accessed 25 January, 2000).

Porter, D. (1997) 'A plague on the borders: HIV, development, and traveling identities in the Golden Triangle', in L. Manderson and M. Jolly (eds) *Sites of Desire, Economies of Pleasure: Sexualities in Asia and the Pacific*, Chicago: University of Chicago Press, pp. 212–232.

Pritchard, A., Morgan, N.J., Sedgely, D. and Jenkins, A. (1998) 'Reaching out to the gay tourist: opportunities and threats in an emerging market segment', *Tourism Management*, 19(3): 273–282.

Pruitt, D. and LaFont, S. (1995) 'For love and money: romance tourism in Jamaica', *Annals of Tourism Research* 22(2): 419–440.

Ralph, E. (2000) Oppose the Trafficking of Women and Children, testimony before the Senate Committee on Foreign Relations Subcommittee on Near Eastern and South Asian Affairs by Regan E. Ralph, Executive Director Women's Rights Division, Human Rights Watch, International Trafficking of Women and Children, 22 February.

Rasmussen, P.K. and Kuhn, L.L. (1977) 'The new masseuse: play for pay', in C. Warren (ed.) *Sexuality: Encounters, Identities and Relationships*, Beverly Hills, CA: Sage Contemporary Social Science Issues, 35, pp. 11–32.

Raymond, J.G. (1995) *Report to the Special Rapporteur on Violence Against Women, The United Nations, Geneva, Switzerland*, North Amherst: Coalition Against Trafficking in Women.

Reuters (1999) 'State employees fuel illicit sex in Vietnam', Reuters (Hanoi), 31 May.

Richter, L.K. (1989) *The Politics of Tourism in Asia*, Honolulu: University of Hawaii Press.

Robins-Mowry, D. (1983) *The Hidden Sun: Women of Modern Japan*, Boulder, CO: Westview Press.

Robinson, G. (1989) 'AIDS fear triggers Thai action', *Asia Travel Trade*, 21 (September): 11.

Robinson, L. (1993) 'In the penile colony: touring Thailand's sex industry', *Nation* 1 November: 492–497.

Rock, P. (1973) *Deviant Behaviour*, London: Hutchinson University Library.

Rogers, C. (1951) *Client Centred Therapy*, Boston: Houghton-Mifflin Co.

Rogers, J.R. (1989) 'Clear links: tourism and child prostitution', *Contours* 4(2): 20–22.

Rojek, C. (1993) *Ways of Escape: Modern Transformations in Leisure and Travel*, London: Macmillan.

Rojek, C. and Urry, J. (1997) *Touring Cultures: Transformations of Travel and Theory*, London: Routledge.

Rose, G. (1993) *Feminism and Geography: The Limits of Geographical Knowledge*, Cambridge: Polity Press.

Rosenfried, S. (1997) 'Global sex slavery', *San Francisco Examiner*, 6 April.

Ryan, C. (1995) 'Conversations in Majorca – the over 55s on holiday', *Tourism Management* 16(3): 207–217.

—— (1997) *The Tourist Experience: A New Introduction*, London: Cassell.

—— (2000) 'Sex tourism: paradigms of confusion', in S. Carter and S. Clift (eds) *Tourism and Sex: Culture, Commerce and Coercion*, London: Cassell, pp. 23–40.

Ryan, C., Hughes, K. and Chirgwin, S. (1999) ' "The gaze", spectacle and eco-tourism', *Annals of Tourism Research* 27(1): 148–163.

Ryan, C. and Kinder, R. (1996a) 'The deviant tourist and the crimogenic place', in A. Pizam and Y. Mansfeld (eds) *Tourism Crime and International Security Issues*, Chichester: Wiley, pp. 23–36.

—— (1996b) 'Sex, tourism and sex tourism: fulfilling similar needs?', *Tourism Management* 17(7): 507–518.

Ryan, C. and Martin, A. (2001) 'Tourist and strippers: liminal theater', *Annals of Tourism Research* 28(1): 140–163.

Ryan, C., Murphy, H. and Kinder, R. (1998) 'The New Zealand sex industry and tourist demand: illuminating liminalities', *Pacific Tourism Review* 1(4): 313–328.

Ryan, C. and Robertson, E. (1997) 'New Zealand student-tourists: risk behaviour and health', in S. Clift and P. Grabowski (eds) *Tourism and Health: Risks, Research and Responses*, London: Pinter, pp. 119–138.

Rybczynski, W. (1991) 'Waiting for the weekend', *The Atlantic Monthly* August: 35–52.

Samuels, A. (1995) 'Gender – a certain confusion', *Achilles Heel* Summer: 10–15.

Sawaengdee, Y. and Isarapakdee, P. (1991) *Ethnographic Study on Long Haul Truck Drivers for Risk of HIV Infection*, Mimeograph. Bangkok: Institute for Population and Social Research, Mahidol University.

Scambler, G. (1997) 'Conspicuous and inconspicuous sex work: the neglect of the ordinary and the mundane', in G. Scambler and A. Scambler (eds) *Rethinking Prostitution: Purchasing Sex in the 1990s*, London: Routledge, pp. 105–120.

Scambler, G. and Scambler, A. (1997) *Rethinking Prostitution: Purchasing Sex in the 1990s*, London: Routledge.

Schechner, R. (1982) *The End of Humanism*, New York: Performing Arts Journal Publications.

Schlessinger, L. (1997) *Ten Stupid Things Men Do to Mess Up Their Lives*, New York: HarperCollins.

Seabrook, J. (1996) *Travels in the Skin Trade: Tourism and the Sex Industry*, London: Pluto Press.

Seebohm, K. (1990) 'A semiotic analysis of the 1990 Sydney gay and lesbian mardi gras', paper presented at the Institute of Australian Geographers Conference, University of New England, Armidale, September.

Selwyn, T. (1993) 'Peter Pan in South-East Asia: views from the brochures', in M. Hitchcock, V.T. King and M.J.G. Parnwell (eds) *Tourism in South-East Asia*, London and New York: Routledge, pp. 117–137.

Sentfleben, W. (1986) 'Tourism, hot spring resorts and sexual entertainment, observations from northern Taiwan – a study in social geography', *Philippine Geographical Journal* 30: 21–41.

Sheehy, G. (1971) *Hustling: Prostitution in our Wide Open Society*, New York: Delacorte.

Shields, R. (1991) *Places on the Margin: Alternative Geographies of Modernity*, London: Routledge.

Shirkie, R. (1982) 'Sex and the simple tourist', *IDRC Reports* 10(4): 6.

Silbert, M.H. and Pines, A.M. (1982) 'Victimization of street prostitutes', *Victimology* 7(1–4): 122–133.

Silver, R. (1993) *The Girl in Scarlet Heels*, London: Century.

Simmel, G. (1971) *On Individuality and Social Forms: Selected Writings*, ed. D.N. Levine, Chicago: University of Chicago Press.

Siren (1998) 'Taxing matters', *Siren (Sex Industry Rights and Education Network), New Zealand Prostitutes Collective* 15: 41–46.

Sisters of the Heart (1997) *The Brothel Bible: The Cathouse Experience*, Las Vegas: Brothel Books.

Skeggs, B. (1994) 'Refusing to be civilized: "race", sexuality and power', in H. Afshar and M. Maynard (eds) *The Dynamics of 'Race' and Gender: Some Feminist Interventions*, London: Taylor and Francis, pp. 106–127.

Skrobanek, S. (1996) 'Foreword', in J. Seabrook *Travels in the Skin Trade: Tourism and the Sex Industry*, London: Pluto Press. pp. vii–ix.

Smith, S.J. and Wilton, D. (1997) 'TSAs and the WTTC/WEFA methodology: different satellites or different planets', *Tourism Economics* 3(3): 249–265.

Solomon, A. (1999) 'Child prostitution starting to appear in Vietnam', Reuters, Hanoi, 11 June.

Sousa, D. (1988) 'Tourism as a religious issue', *Contours* 3(5): 5–13.

Spectrum (1998), Radio New Zealand, 13 October.

Sprinkle, A.M. and Gates, K. (1997) *Forty Reasons Why Whores Are My Heroes*, from vol. 1, XXX000, New York: Gates of Heck Inc, also http://www.heck.com/annie/40reasons.htm (accessed March 1998).

Stark, E. and Flitcraft, A. (1996) *Women at Risk: Domestic Violence and Women's Health*, Thousand Oaks, CA: Sage Publications.

Stokes, J. (1994) 'Prudes on the prowl: the view from the Empire Promenade', in G.M. Young (ed.) *Victorian England*, London: The Folio Society, pp. 365–396.

Stoler, A. (1997) 'Educating desire in Colonial Southeast Asia: Foucault, Freud and imperial sexualities', in L. Manderson and M. Jolly (eds) *Sites of Desire, Economies of Pleasure: Sexualities in Asia and the Pacific*, Chicago: University of Chicago Press, pp. 27–47.

Stoller, R.J. (1975) *Perversion: The Erotic Form of Hatred*, New York: Pantheon Books.

—— (1979) *Sexual Excitement: Dynamics of Erotic Life*, New York: Pantheon Books.

—— (1991) *Porn: Myths for the Twentieth Century*, New Haven, CT: Yale University Press.

Sullivan, B. (1997) *The Politics of Sex: Prostitution and Pornography in Australia Since 1945*, Cambridge: Cambridge University Press.

Sutton-Smith, B. (1972) 'Games of order and disorder', paper presented to Symposium on 'Forms of Symbolic Inversion', American Anthropological Association, Toronto, 1 December.

Swinglehurst, E. (1982) *Cook's Tours: The Story of Popular Travel*, Poole: Blandford Press.

Sydney Morning Herald (1995) 'Ending the sex tours', *Sydney Morning Herald*, 23 June: 17.

—— (1999) 'Column 8', *Sydney Morning Herald*, 11 February.

Symanski, R. (1981) *The Immoral Landscape: Female Prostitution in Western Societies*, Toronto: Butterworths.

Thane, P. (1999) 'Late Victorian women', in G.M. Young (ed.) *Victorian England*, London: The Folio Society, pp 335–364.

Thanh-Dam, T. (1983) 'The dynamics of sex-tourism: The cases of Southeast Asia', *Development and Change* 14(4): 533–553.

Theweleit, K. (1987) *Male Fantasies, Vol. 1, Women, Floods, Bodies and History*, Cambridge: Polity Press.

Thomson, A. (1996) 'Political mistresses reveal state of affairs', *Dominion*, 3 April: 7.

Tonry, M. and Morris, N. (1985) *Crime and Justice*, Chicago: University of Chicago Press.

Truong, T-D. (1983) 'The dynamics of sex-tourism: the cases of Southeast Asia', *Development and Change* 14(4): 533–553.

—— (1990) *Sex, Money and Morality: Prostitution and Tourism in South-East Asia*, London: Zed Books.

Tsoulis, A. (1996) *I'll Make You Happy*, A Blue Angel Films Release, Auckland, New Zealand.

Turner, R. and Surace, S.J., (1956) 'Zoosuiters and Mexicans: symbols in crowd behavior', *American Journal of Sociology* 62(1): 14–20.

Turner, V. (1969) *The Ritual Process – Structure and Anti-structure*, London: Routledge and Kegan Paul.

—— (1974) *Dramas, Fields and Metaphors: Symbolic Action in Human Society*, Ithaca, NY and London: Cornell University Press.

—— (1982) *From Ritual to Theater: The Human Seriousness of Play*, New York: PAJ Publications.

TVNZ News (1999) 'Hero Parade: biggest ever crowd', *TVNZ News*, 14 February.

UNESCO and Coalition Against Trafficking in Woman (1992) *The Penn State Report: International Meeting of Experts on Sexual Exploitation, Violence and Prostitution*, Pennsylvania: State College.

Urry, J. (1990) *The Tourist Gaze*, London: Sage.

US Congress (1999) *Bill Summary and Status for the 106th Congress*, (introduced 3 November), http://thomas.loc.gov/cgi-bin/bdquery/z?d106:SN00600:@@@D&summ2=m& (accessed 25 January, 2000).

Wagner, U. (1977) 'Out of time and place – mass tourism and charter trips', *Ethnos* 42: 38–52.

Watkin, H. (1999) 'Victim exposes sex-slave gang', *South China Morning Post*, 29 July.

—— (2000) 'Denial and hypocrisy fuel growing sex trade', *South China Morning Post*, 4 March.

Wherrett, R. (ed.) (1999) *Twenty-one this Year, the Sequinned Revolution Earns a History 'Mardi Gras! True Stories: From Lock Up to Frock Up'*, Melbourne: Viking.

Wickens, E. (1994) 'Consumption of the authentic: the hedonistic tourist in Greece', in A.V. Seaton, C.L. Jenkins, R.C. Wood, P.U.C. Dieke, M.M. Bennett, L.R. MacLellan and R. Smith (eds) *Tourism: The State of the Art*, Chichester: Wiley, pp. 818–825.

—— (1997) 'Licensed for thrill: risk taking and tourism', in S. Clift and P. Grabowski (eds) *Tourism and Health: Risks, Research and Responses*, London: Pinter.

Wihtol, R. (1982) 'Hospitality girls in the Manila tourist belt', *Philippine Journal of Industrial Relations* 4(1–2): 18–42.

Wilson, D. (1997) 'Paradoxes of tourism in Goa', *Annals of Tourism Research* 24: 52–75.

Winter, M. (1976) *Prostitution in Australia*, Balgowlah: Purtaboi Publications.

WTTC (1997) 'Rejoinder: from the World Travel and Tourism Council', *Tourism Economics* 3(3): 282–288.

Yamba, B. (1988) 'Swedish women and the Gambia', in T. Selwyn (ed.) *Conference on the Anthropology of Tourism*, London: Froebel College.

Yeatman, A. (1990) 'A feminist theory of social differentiation', in L.J. Nicholson (ed.) *Feminism/Postmodernism*, New York: Routledge.

Yeo, A. (1999) *Subject: Mardi Gras is fucked*, (email) 16 February.

Yiannakis, A. and Gibson, H. (1992) 'Roles tourists play', *Annals of Tourism Research* 19(2): 287–303.

Yiu, Y. (1994) 'It's the fashion to have a concubine', *Next*, 2 September: 72–73.

Index